LAWS:
RIGIDITY
AND
DYNAMICS

LAWS:
RIGIDITY
AND
DYNAMICS

Edited by

Eliezer Rabinovici
The Hebrew University of Jerusalem, Israel

World Scientific

NEW JERSEY · LONDON · SINGAPORE · BEIJING · SHANGHAI · HONG KONG · TAIPEI · CHENNAI · TOKYO

Published by

World Scientific Publishing Co. Pte. Ltd.
5 Toh Tuck Link, Singapore 596224
USA office: 27 Warren Street, Suite 401-402, Hackensack, NJ 07601
UK office: 57 Shelton Street, Covent Garden, London WC2H 9HE

British Library Cataloguing-in-Publication Data
A catalogue record for this book is available from the British Library.

Cover images:
Ancient Chinese Laws 'Shujing' (Source: Wikipedia); Ten Commandments (Source: Shutterstock);
Physics Laws 'Standard Model Formula' (Source: CERN); Superstition 'Black Cat' (Source:
Shutterstock); Hammurabi's Law Code (Author: Deror avi); Law of Demand in Economy
(Source: Wikipedia); Mendelian Genetics 'Punnett Square Mendel Flowers' (Author: Madprime).

LAWS: RIGIDITY AND DYNAMICS

ISBN 978-981-12-9230-9 (hardcover)
ISBN 978-981-12-9231-6 (ebook for institutions)
ISBN 978-981-12-9232-3 (ebook for individuals)

For any available supplementary material, please visit
https://www.worldscientific.com/worldscibooks/10.1142/13816#t=suppl

Typeset by Stallion Press
Email: enquiries@stallionpress.com

Contents

Introduction

The Inter-Continental Academia:
An Adventure in Progress

Eliezer Rabinovici

This volume presents examples of what transpired before, during, and after the third Inter-Continental Academia (ICA3) events held mainly at Nanyang Techno-logical University (NTU) Singapore and the Institute of Advanced Studies (IAS) at the University of Birmingham UK, during 2018 and 2019. Two interim events were held in the fall of 2018 at the IAS Princeton and the IAS of the Techni-cal University of Munich. ICA is held under the auspices of the network of University-Based Institutes of Advanced Studies (UBIAS). The theme of ICA3 was Laws: Rigidity and Dynamics. This chapter gives the background upon which the ICA concept was conceived and the vision it wishes to promote, as well as an overview of the contributions to the ICA3 meetings. I began writing before the Covid virus caused a pandemic and am concluding writing it before it is clear to what extent we are out of the woods. The course of events has left an imprint on this chapter.

To Lars Brink, an insightful and loyal travel
companion in many adventures.

Loneliness, Vision and Personal Experience

Founding the ICA is an act towards strengthening unity and the commu-nity. According to the scriptures, humanity was banished from paradise and dispersed to somewhere else on the planet. Many of the challenges that life poses, from the sweat needed for providing food to the pain as-sociated with birth, were attributed to that primordial traumatic event in human existence. However, let's admit it, life even outside paradise, but anywhere on Earth, is by far a superior paradise once one considers the alternatives. The physical conditions in any other place in the solar system and for light years away from us are incredibly harsh. Life is not welcome there. The same holds when one attempts to leave the thin crust of Earth and dig underneath it; we as living beings are isolated and alone.

The loneliness felt by the first humans venturing outside Africa pales in comparison to the loneliness of the human race confined to the surface of our planet. This has a fundamental effect on humanity as a whole and is also an essential ingredient of the individual human condition. One does not need to have the highly sensitive soul of a poet for us to feel this. How does humanity react to this, how do humans react to this?

Visible, concrete, heroic, and somewhat futile attempts involve physically trying to extend our sphere of existence in space and in time. Constructing tall buildings, ships, balloons, airplanes, space rockets, submarines and deep space probes is one way to try and defy our condition. Discovering ways to increase the quality and the longevity of life is another. This is a very challenging and painfully slow process in which only very few are engaged. The social responses are polarized. At one pole, humans form communities and humanity lays foundations for uniting nations. Around the second, one constructs shrines of all-powerful individuals and nationalism. The tension between these poles brings out the best and the worst in us as humans and as nations. It also led me to originate the vision of an ICA, this in a hot room in the New Delhi IAS during a meeting of some University Based Institutes for Advanced Studies (UBIAS) directors.

Before diving into the ICA concept and history in some detail, I will reflect on my personal scientific experience of this struggle. I am a physicist and my area of research is mainly of theoretical high-energy physics. We as humans are in addition confined to a body that extends from a few centimeters to at most a couple of meters and yet we believe that somehow, we are owed a concise explanation of the whole universe and the matter it consists of. These demands include the understanding of what are the laws and structures ruling the behavior of matter all the way from distance scales that are incredibly smaller and unimaginably larger than we are. Of course, we are not owed this and we are hard wired to ignore this NOH (Nobody Owes Humanity) principle with typical human hubris. Our superiority and inferiority feelings fight each other. Some react to this by wearing t-shirts from their twenties well into their seventies, others hide behind the smoke of high-class pipes. Socially, some isolate themselves and others prefer teamwork, some are forced, willingly or not, by the very magnitude of the problem into collaborations. Intellectually, in the effort to be able to compress all the knowledge into one part of the brain, scientists in this field have been striving for decades to find a Grand Unified Theory to explain in a concise manner all the basic laws and elementary constituents of matter. This gallant vision has achieved various degrees of success, some very impressive, over the years. Yet, those very people who have the privilege to stare at this panorama and comprehend pieces of it have also become experts in hair splitting when it comes to deciding how to judge the work of their fellow scientists. To whom will they decide to assign credit and, not less important, from whom will they decide to remove it.

Back to the room in the IAS in New Delhi and the ICA idea.

IASs and UBIASs

The foundations of the ICA rests on the UBIAS.

The idea of constructing a community of scholars extends back to the Middle Ages. A precise concept of what should be the desired composition of such a community evolved over the centuries. In the 20th century the founding of an IAS in Princeton in 1930, established a benchmark of excellence.

It was constructed to consist of faculty whose work is of the highest scientific quality. The primary mission of these faculty members, who held permanent positions, was defined to mentor the post-graduate fellows who were invited to spend several years at the IAS. University students were not supposed to be part of the community. Following many aspects of the Princeton model, IASs have been established in Berlin, Palo Alto, Harvard, Amsterdam, Jerusalem, North Carolina and Uppsala to mention a few. Some, but not all, of these institutes were founded as independent entities. Others had ab-initio strong links with the University in which they resided. The latter are denoted by University-Based Institutes of Advanced Studies (UBIAS). They are united by their dedication to the advancement of knowledge and diverge in the manner in which they try to achieve it. The comparison between UBIASs and IIASs (Independent IASs) is interesting and instructive; it deserves a special treatment which I will not attempt here. I will note, however, that in 2010 the UBIAS network was founded when representatives from 32 research institutes worldwide met at the conference entitled *"University-Based Institutes for Advanced Study in a Global Perspective: Promises, Challenges, New Frontiers"*. The meeting was initiated and hosted by the Freiburg Institute of Advanced Studies, Germany. A volcanic eruption in Iceland caused several months of delay in convening the meeting and was a clear signal to the interconnectivity of the planet. The network now consists of over fifty members. A small steering committee, of which I became a member, was established.

At the time, I was the director of the IAS in Jerusalem. We met in March 2012 at the Jawaharlal Nehru Institute for Advanced Study, in New Delhi, India. We were searching for something which would distinguish the UBIAS network from other IASs. I considered that the main and unique asset of UBIAS is that its members are situated on all continents (save Antarctica).

I had a rather ambitious vision that one could thus use the UBIAS as a substrate on which to grow gradually a network of high-quality academics coming from all continents at various age levels. This would evolve into a cadre of outstanding global, multidisciplinary scholars who have learned to work together and respect each other's viewpoints. A pool of extraordinary academics who may find ways to serve, analyze, and propose solutions to global challenges. The tool I suggested to accomplish this was an Inter-Continental Academia (ICA), this brand name was proposed by Martin Grossman and Carsten Dose.

ICA Concept

The ICA would convene for about ten days one year at an IAS on one continent and the next year at a different IAS on another continent. An intellectual theme would be defined for both meetings. The ICA itself would be comprised of highly accomplished mentors and early career, promising fellows. As is generously described on the UBIAS website: "The initial idea was of the then director of the Israel IAS Professor Eliezer Rabinovici. The IEA-USP and the IAR (Institute for Advanced Research) of the Nagoya University felt motivated by the insight of the Israeli scientist and decided to be behind the pilot project for this initiative, which operated as a joint venture."

Mission and Motivation

The mission is ambitious, the problems facing those living on planet Earth in the 21st century are becoming more and more global in nature and quite formidable, at least for those people and institutes who preside over, democratically elected or not, nations as well as large companies. One may even be concerned that these issues are becoming too complex for most people. Neither the individual citizens nor their leaders are equipped to deal with them effectively. The process of the transfer of power in response to complexity has been evolving for thousands of years. From individuals to tribes, urban centers, countries, EU and corporations. This is out of our hands as academics to influence it. Nevertheless, let's anyhow take a step in that direction, as small as it may seem. Recognizing the intercultural and intercontinental reach of the challenges, let us try to lend a hand to face them appropriately. The very ambitious idea is to gradually build up a network of outstanding multidisciplinary scholars from across the globe who have learned to work together and respect each other's viewpoints. A pool of academics, at various stages of their career who may find ways to serve, analyze, and propose solutions to global challenges. The pandemic intensified again the tension between contracting to the minimal social cells for safety, and the longer-range need to pull in all possible resources to combat the virus. While travel has become both much more complicated and environmentally threatening, it has become even more important to learn how to bridge cultural gaps that become walls in the face of a crisis that carries strong emotional components. This understanding and trust building requires face-to-face interactions. Luckily, our planet is small enough to allow for them.

Mentors and Fellows: Who should they be and what does one expect from them?

Mentors: The goal is to strive to involve leaders in their field while aiming for an intercontinental coverage as well as an appropriate gender balance. The mentors should be willing to be adventurous, willing to interact with each other and with the fellows, and be curious because they will be jumping into the unknown. This has been achieved quite reasonably in different ICAs. The complete list demonstrating this is in the appendix.

ICA 3 Laws: Rigidity and Dynamics. There were several expectations of the mentors. First to participate for as much time as is possible for them. To present their point of view of the subject of the meeting, to have formal and informal interactions with the fellows, and supply the fellows with feedback when appropriate. They are also expected to serve as living examples that authority resides in the merit of ideas and not the decorations bestowed on their originators. That is the only thing that counts in academic discourse. Ideas presented by Nobel Laureates are challenged with the same vehemence as those presented by the fellows. The mentors should allow themselves to be playful when different ideas and personalities present themselves.

Fellows: One should aim for the same diversity profile as that of the mentors. The goal is to involve as fellows those who have a significant chance of remaining within the academic system. The rank we aim at is mostly for the fellows at the level of assistant professors. At the later stage of their postdoctoral fellowship and as assistant professors, they tend to be preoccupied with firming up their tenure positions and are usually less free to form bonds. The ICA allows them to open their vistas when they are best equipped to do so and least inclined to risk it.

The choice of the fellows should respect the intercontinental range in full as well as a gender balance.

A word here on gender balance is in order, this is in recognition that all genders are invaluable sources of human talent and in this time and age; all such resources should be acknowledged and fully tapped.

The fellows should be rising scholars in their field with a mind open to appreciate and respect very different approaches to a subject they thought they had mastered. They should be willing to learn from different disciplines using methods less familiar to them. They should have the skills needed to identify the potential of the knowledge that becomes available to them. This can go all the way from growth in their understanding of a subject, through formulating joint problems of interest, requesting joint funding for concrete new research projects growing out of the ICA, right up to taking a totally different career path after the ICA. All of these options have actually been outcomes of participating in the ICAs.

This was the original concept and it has evolved from the original ICA 1 to the current ICA 4, but oops, I forgot one component, how to define the all-important, for some, metrics for success. Joint research, joint research proposals, articles and books, etc. The reason for that is that for me then, as well as essentially now, the main idea was to weave the network. The rest would come out when it should and naturally. Casting your bread upon the waters in this way is a long-range investment, which should be nurtured and can be evaluated in due time. Additional outputs reflect the different participants and the priorities of individual host institutes.

From Vision to Reality

Two UBIAS institutes each on a different continent declare to the UBIAS members their interest to co-host an ICA. This happened in New Delhi where Nagoya and São Paulo jumped into the water. The directors of the institutes with the advice of their respective pool of consultants chooses a challenging topic with multidisciplinary aspects. For ICA1 that was on the various aspects of time. The next step is to identify and invite high quality mentors to join the adventure. As mentioned, the group should be diverse in their continental and intellectual interests and an effort should be made to assure the best possible gender balance. Their number may range from approximately ten to fifteen. Once the group is formed, UBIAS advertises the ICA's program to all its members and asks them to consider an active role in the next step, to identify and nominate appropriate fellows. In addition, the hosts of the ICA search for candidates to be fellows by other means. This is not ideal but what is possible at this stage. The directors should be committed to cover the travel expenses of the candidates they have nominated if those have been elected. The number of fellows chosen can also range approximately from ten to twenty. One should target that the team of fellows have a similar profile as that of the mentors in terms of diversity. The actual selection process for ICA3 as a case study will be described by Sue Gilligan in her contribution to this volume.

The Program

One of the challenges which is very rewarding to overcome is constructing a program acceptable to both hosts. A large amount of energy was released when the directors of the IASs in Nagoya and São Paulo were preparing a

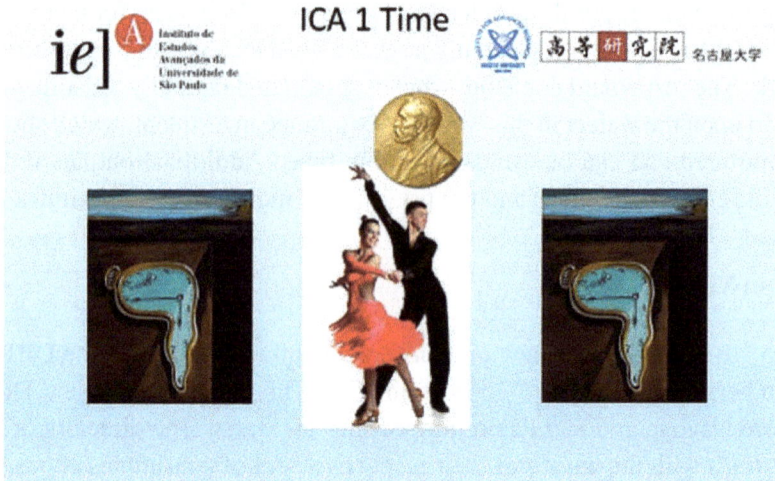

fusion menu. Very superficially, it could be considered as the Nobel Prize Medal meets the Samba Dance.

The lists of mentors, fellows and organisers of the programs appear in Figs. 1–5. Figures 6–9 are photos of the participants, and on Pages xxxviii– xl one finds the detailed programs of the two meetings they were held in São Paulo, Brazil and then in Nagoya, Japan. The mentors and fellows provided a high-quality scientific program on each continent and were also acquainted with the respective cultures. They visited museums and hospitals in addition, were exposed to the challenges that large urban areas present to their inhabitants. The fellows participating in ICA 1 have remained in touch from the first meeting until this very day. At the time this book was completed the fellows came out with an amazing MOOC Off the Clock: The Many Faces of Time.

The topic of ICA 2 was Human Dignity. It was held at the IIAS Jerusalem, Israel and at Bielefeld, Germany and it had a follow up meeting at Johannesburg, South Africa. The lists and photos of participants, organisers of the meetings appear in Figs. 1–5, 10–12, and the programs are on Pages xli–xliv and xlv–xlvii.

ICA 3 — Nuts and Bolts

The IAS of Birmingham and the NTU Singapore volunteered to host ICA 3. The two continents involved in general and the two countries in particular had seen the nature of their relations evolve over the years. The director on behalf of the NTU was myself and the director on behalf of the IAS of Birmingham was Professor Michael Hannon.

The topic *Law: Rigidity and Dynamics* was selected to provide a forum and framework for the intellectual exchange. The concept of laws has different meanings in different settings, cultures and to individuals. Yet "laws" are common from Medicine to Linguistics, Chemistry to Courts, History to Physics, Psychology to Economics, and Engineering to Theatre.

This rich theme demanded reflection, analysis and comparison, to elucidate the origins of laws and the tensions they create. When, why, how and in what cultures are they discovered or dictated? When and why are they broken? How is that common, or different, across disciplines and why? Whether examining laws devised by humans, nature, providence or those evolving and emerging from chaos and complexity one finds oneself examining the limits of the freedom humans have.

The potential of this explosive cocktail attracted illustrious mentors, all indeed leaders in their respective fields. The knowledge and the diversity they share among them and brought to the program is evident from the list on Pages xlviii–lv.

The process of the selection of the fellows is described in detail in the chapter by Ms. Sue Gilligan and Prof. Kwek Leong Chuan (Chapter 12 of the present book). The list appear in Fig. 5.

The group assembled from all over to Singapore. On March 18, 2018 the ICA started.

It commenced along the lines of an experimental concept for the program was set up allowing for an in-situ evolution.

The concept was to start off the program by having the fellows introduce themselves and their research interests to each other. This was followed by several days on which the emphasis was on presentations by mentors and the fellows. Each mentor presentation was responded to by another mentor from a field afar of that of the presenting mentor, as well as fellows. The request was that the responders challenge, if relevant, some points of the presentation. One motivation for that was to remove from the fellows a possible reluctance to question assertions made by giants in the fields such as Nobel Laureates. This worked out quite well, the prevailing atmosphere was of very open discussions. The presentations by the fellows allowed the other fellows and the mentors to better understand the intellectual background from which they were coming. The presentation sessions climaxed on a date in which one managed to have almost the whole group of mentors being present. As the time passed the leadership was passed on to the fellows, they were expected to and they did set up the program. As they were absorbing the vast and fascinating panorama that the theme of Law presented to them, they were gradually trying to identify research projects that would be of common interest to them. This was not a smooth process, but it was interesting. During the meeting all participants were allowed to experience many various unique aspects of Singapore, nature, culture and history. The detailed culture and intellectual program in Singapore are on Pages xlviii–lv. Highlights of the cultural programs in Singapore and Birmingham are described in more detail in Chapter 12 by Sue Gilligan and Kwek Leong Chuan.

As the Singapore leg of the meeting was reaching conclusion the fellows were given another fantastic opportunity. Patrick Geary and Ernst Rank very generously invited fellows to the IAS Princeton, USA and the IAS in TUM, Germany to develop further their common research ideas before the Birmingham leg and under their mentorship.

These stepping-stones meetings did indeed happen in Princeton October 4–8, 2018 with the topic *'Deviating from Laws'* and at TUM Munich October 26–27, 2018 with the topic *'Changing Laws'*.

Almost exactly a year later, on March 19, 2019, ICA 3 reconvened in Birmingham. Most of the mentors arrived for this second leg attesting to the interest they had found in the program. All the fellows, but one, returned as well. In addition, new mentors had arrived as well. The detailed scientific program and cultural program appears on Pages lvi–lx. During this leg a large part of the schedule was directed by the fellows themselves, they were making progress in defining a common path. This was done with considerable feedback provided by mentors. Overall, the level of the interactions and presentations was very high. I will be describing later and in some detail those which resulted in contributions to this volume.

Returning for a moment to the UBIAS aspect, ICA 3 was integrated in an enriching and exemplary manner in the University of Birmingham activities. Sue Gilligan will describe in her chapter with Kwek Leong Chuan how this was done. On March 27th a major part of ICA 3 was brought to a conclusion, plans for the next steps were drawn and the Inter-Continental convergence was replaced by a Inter-Continental dispersal. Little did we guess how the freedom of travel we took for granted will be curtailed in the coming years.

Outputs and Further Steps

The ICA is an adventure in progress. I will describe some intermediate conclusions on the success the challenges it faces and what I believe should be done allow it to flourish. One may also recall that the project is at the stage of a flower bud waiting to blossom.

Positive lessons

- The host institutes were all extremely dedicated and managed to raise the necessary funds to accommodate the fellows and mentors.
- The topics chosen all had depth and were very interesting.
- All the ICAs succeeded to attract an inter-continental and multi-disciplinary diversity of world leading academics. Their active participation was intense.
- The fellows of all ICAs were highly interacting, devoted and formed an inter-continental rainbow.
- The feedback from both mentors and fellows were very encouraging.

Need to improve

- The burden to raise the financial resources needed for running and ICA and involved in recruiting of fellows fell mainly on the host institutes. It is my opinion that is the Achilles heel of the project. The ICA should either find long term sponsors or be better integrated within the UBIAS.

- Concretely ICAs need a solid funding and to increase the pool of fellows that apply. ICA needs to be able to brand itself. The first steps have been taken.

ICA 4 Intelligence and Artificial Intelligence — an ICA in the Making

It is crucial that fellows and mentors conceive their ICA as one link in a chain of excellence. Thus while ICA 3 was in full swing the idea of ICA 4 was starting to crystalize. The first initiative came from the IAS Marseilles to collaborate with the IAS Belo-Horizonte on the theme of Intelligence and Artificial-Intelligence. With such a topical theme, arranging for it seemed to be a shoe in. The spread of Covid-19 changed all the rules and yet with amazing persistence of the directors outstanding mentors and fellows were recruited and a preliminary stage zero has already taken place with all the gadgets that zoom like meetings have brought.

Postscript: Well after all contributions to this book have arrived, ICA 4 happened and was concluded. The IAS Paris took the helm from IAS Marseilles. The meetings had indeed been held in person in both Paris (2021) and Belo-Horizonte (2022). With Interim meetings in the IAS Birmingham and Nagoya. The academic level of the mentors and the fellows remaining very high.

Alumni

To fulfill the mission of creating a network of Inter-Continental fellow and mentors we started building up an Alumni Network. This also to instill an esprit de corps in all fellows. One asset that sprung alongside the many liabilities that Covid has brought with it was the ability to have low budget large meetings. We had a zoom meeting in which most the fellows of ICAs 1, 2 and 3 participated. They had shared with each other the various benefits they drew from their attending the ICAs. Many of the mentors of ICA3 also attended the zoom meeting and described their own experiences. This is a first and promising step in realizing the idea of an Inter-Continental network.

I described the brief history of the ICA, it is an adventure in progress, the range of participants involved is indeed intercontinental. The project should be viewed as a long-range investment which must be nurtured and can be reevaluated in due time. Like all adventures, it may or may not end up having an important impact and one will never know without trying.

The Book

The final section is an introduction to the content of this book.

Penelope Andrews was the Dean of the Law School in Cape Town when she participated in the ICA. She is now in the New York Law School. She is involved in the most direct meaning of the Law. Her chapter **The "Casserole" Constitution: The South African Constitution and International Law** describes and analyses, as a first-hand witness, the dynamics that can allow Laws dictated by humans to evolve. The experience in South Africa is one of the most positive, intense and revolutionary that has ever taken place. It is described in depth and is a source of inspiration.

The four following chapters are united, not only by their overt and latent passions but also by the interplay between History and the Laws of Physics they address.

Patrick Geary, a Historian from the IAS Princeton, has discussed **Laws of History and Laws in History**. Patrick was the most dedicated and appreciated mentor by the fellows. He shares in the chapter his deep insights and knowledge in the history of laws on the one hand, while he constructs a deep and fortified moat to protect history from any attempts to attribute universal laws to it.

David Gross, a Physicist from UC Santa Barbara has titled his contribution as **The Nature of Laws and Principles in Science**. A competent future historian will easily notice that this chapter, championing the scientific method as a tool to define truth, was written under the influence of the turbulent Trump years. David makes a passionate and firmly grounded case for a future where the scientific method and the laws it uncovers is a key navigating tool for humanity.

Lars Brink, a Physicist from Goteborg, bears some responsibility for some of the Nobel Prizes awarded. He has contributed the chapter **Are there Physics Laws or are They Like Skins of Onions**. He describes how some physicists addressed this question being sensitive both to the transient aspects of laws of Physics as well as their potential trespassing on religious grounds.

Michel Spiro an experimental Physicist from the University of Orsay and Maurizio Bona from CERN contributed the chapter **The CERN Model, a Collective Machinery to Test Laws Against Nature**. They describe the experience of CERN both as a model to conduct Inter-Continental collaborative scientific research at the highest scientific and human level. A very concrete, extraordinary and explicit example of how the scientific method is used to discover the laws of nature on a global scale.

Gil Kalai, a Mathematician from the Hebrew University, Jerusalem has written the chapter **The Argument against Quantum Computers, the Quantum Laws of Nature, and Google's Supremacy Claims**. Gil, a David against Goliaths, explains in detail what are classical and quantum computers. He elaborates on how the laws of nature, which are quantum, will be an insurmountable obstacle to any effort to construct a quantum computer. Technological giants are employing outstanding scientists and investing vast amounts of funds to construct these very quantum computers.

Partha Dasgupta is an Economist from Cambridge UK, has written the chapter **Laws and Norms as Social Institutions** . It brings us to the realm of the social sciences. Partha paints the role of institutions in economic life. By the title he has chosen he meant, very loosely, the fundamental *arrangements* that govern collective human undertakings. From families to firms in countries with a variety of economical levels.

Martin Freer, a Physicist from Birmingham a hub of the industrial revolution has contributed a chapter titled **Complexity in Energy and a Low Carbon Transition**, where he discusses the pros and cons of each of the choices on where to concentrate the efforts to reach zero net carbon emission. The national scale intervention via market driven initiatives and the providing local energy solutions are the two choices he analyses.

William J. Chaplin in his chapter **Physics Meets Art at the University of Birmingham** describes a fascinating multidisciplinary project he was involved with. Artists and scientists, in particular physicists, meeting to create an enlarged space together.

The all-important closing chapters were written by the fellows themselves.

Alastair Wilson dealt with his different thoughts about the exemplary role the laws in Physics should play in the pursuit of knowledge in general, this in his chapter **Laws about Laws**. He embraces the important role these laws have in explaining some of the natural phenomena and he does not accept them as an omnipotent desired framework to describe all aspects of knowledge. Based also on his experiences in the ICA3 meetings he suggests to consider more general and more unifying concepts of Law.

A group of fellows consisting of Hanjo Hamann, Ulrich Heisserer, Nkatha Kabira, Isabel Kusche, Irina Kuznetsova, Petra Liedl, Tom Schonberg and Carla Aparecida Arena Ventura have chosen for their common work the topic **Dynamic Cities and Rigid Laws?** This was not a presentation done during the ICA, it describes the fruit of their common research during and after the official ICA gatherings. This chapter explores the potential and limits of laws to improve livability in cities. This is chapter written by futures leaders in their respective fields on the problems and challenges of the present and the future.

Admin: The real hard work of holding an ICA falls upon the administrators. . . .

Eliezer Rabinovici, a Physicist from the Hebrew University, Jerusalem, myself, describes from a personal viewpoint how the idea of the ICA was conceived, actualized and evolves. In the chapter called **The Inter-Continental Academia: An Adventure in Progress** I describe its short history, its mission and how I view its future prospects.

Thanks

Thanks go to all the ICA host institutes, fellows and mentors.

I would like to give personal special thanks to members of the ICA 3 organizing committee: Kwek Leong Chuan, Lars Brink, Sue Gilligan, Michael J Hannon and Phua Kok Khoo.

In addition I would like to thank those involved in the original ICA 1: Martin Grossman, Hisanori Shinohara, Depeng Cai and in the latest ICA 4: Raouf Boucekkine, Saadi Lahlou, Estevam Barbosa de Las Casas, Olivier Bouin, Sue Gilligan and Guilherme Ary Plonski.

Without you and your dedication the ICA idea could not have been tested. Finally, thanks to the team at World Scientific Singapore and especially Lim Chee Hok for preparing this book for publication.

Lists of Participants and Programmes

Organisers			
ICA 1 *Time*			
Carsten Dose	General Secretary	University of Freiburg	Europe
Dapeng Cai	Economics	University of Nagoya	Asia
Martin Grossmann	Fine Art	University of São Paulo	South America
Regina Markus	Chronopharmacology	University of São Paulo	South America
Takao Kondo	Biological science	Nagoya University	Asia
Bernd Kortmann	Linguistics	University of Freiburg	Europe
Sami Pihlström	Philosophy	University of Helsinki	Europe
Eliezer Rabinovici	Particle Physics	Hebrew University in Jerusalem	Asia
Till Roenneberg	Chronobiology	Ludwig-Maximilians University München	Europe
Hisanori Shinohara	Molecular Science	Nagoya University	Asia
ICA 2 *Human Dignity*			
Ulrike Davy	Law	Zentrum für interdisziplinäre Forschung	Europe
Alon Harel	Law	Hebrew University of Jerusalem	Asia
Hanna Lerner	Political Science	Tel Aviv University	Asia
Michal Linial	Molecular Biology	Hebrew University of Jerusalem	Asia
Benny Porat	Law	Hebrew University of Jerusalem	Asia
Nadiv Mordechay	Project co-ordinator	Israel Institute for Advanced Studies	Europe
Marc Schalenberg, ZiF	Project co-ordinator	Zentrum für interdisziplinäre Forschung	Asia
ICA 3 *Laws: Rigidity and Dynamics*			
Lars Brink	Physics	Chalmers University of Technology	Europe
Kwek Leong Chuan	Quantum Technology	National University of Singapore	Asia
Fiona de Londras	Law	University of Birmingham	Europe
David Gange	History	University of Birmingham	Europe
Sue Gilligan	Project co-ordinator	University of Birmingham	Europe
Michael J Hannon	Chemistry	University of Birmingham	Europe
Phua Kok Khoo	Physics	Nanyang Technological University	Asia
Robin Mason	Economics	University of Birmingham	Europe
Eliezer Rabinovici	Particle Physics	Hebrew University in Jerusalem	Asia
Jonathan Reinarz	History of Medicine	University of Birmingham	Europe
Damien Walmsley	Dentistry	University of Birmingham	Europe

Fig. 1.: Organisers.

Eliezer Rabinovici

Mentors and Speakers

ICA 1 *Time*			
Fernando Aith	Medicine	University of São Paulo	South America
Neka Menna Barreto	Gastronomy	University of São Paulo	South America
Tiago Bosisio Quental	Ecology	University of São Paulo	South America
Ruud Buijs	Neurobiology	Nacional Autónoma de México	South America
Massimo Canevacci	Anthropology	Università degli Studi di Roma La Sapienza,	Europe
Sylvia Duarte Dantas	Psychology	São Paulo Federal University	South America
Carolina Escobar	Anatomy	Nacional Autónoma de México	South America
Luiz Gylvan Meira Filho	Climate Change	UN Climate Convention.	South America
Karl-Heinz Kohl	Anthropology	Goethe-University Frankfurt	Europe
Fernando Iazzetta	Music	University of São Paulo	South America
Vera Lúcia Imperatriz-Fonseca	Biosciences	University of São Paulo	South America
Matthew Kleban	Theoretical Physics	New York University	North America
Takao Kondo	Biological Science	Nagoya University	Asia
Hideyo Kunieda	Physics	Nagoya University	Asia
Eduardo Monteiro	Music	University of São Paulo	South America
René Nome	Chemistry	State University of Campinas	South America
Leopold Nosek	Psychiatry	University of São Paulo	South America
Rodrigo Oliveira	Chef	Mocotó's restaurant	South America
Suzana Pasternak	Architecture	University of São Paulo	South America
Renato Janine Ribeiro	Ethics	Brazilian Minister of Education	South America
Ana Lydia Sawaya	Physiology	São Paulo Federal University	South America
Paulo Saldiva	Medicine	University of São Paulo	South America
Laymert Garcia dos Santos	Sociology	University of São Paulo	South America
Hugo Segawa	Art	University of São Paulo	South America

Fig. 2.: Mentors and Speakers (ICA 1).

Mentors and Speakers

ICA 2 *Human Dignity*			
Aleida Assmann	English Literature	University of Konstanz	Europe
Admiral Ami Ayalon	Political Science	Former director of Israel Security Agency	Asia
Aharon Barak	Law	Radzyner Law School and Yale	North America
Ben-Sasson	Jewish History	Hebrew University of Jerusalem	Asia
Meir Brezis	Medicine	Hebrew University of Jerusalem	Asia
Bernadette Brooten	Christian and Gender Studies and Classics	Brandeis University	North America
Marcus Duwell	Philosophy	Utrecht University	Europe
Moshe Halbertal	Philosophy	Hebrew University of Jerusalem;	Asia
Christine Hayes	Religious Studies	Yale	North America
Lynn A. Hunt	History	University of California, Los Angeles	North America
Michael Karayanni	Law	Hebrew University of Jerusalem	Asia
Mordechai Kremnitzer	Law	Hebrew University of Jerusalem	Asia
Gertrude Lübbe-Wolff	Law	University of Bielefeld	Europe
Raya Morag	Cinema Studies	Hebrew University of Jerusalem	Asia
Sari Nusseibeh	Philosophy	Al-Quds University in Jerusalem	Asia
Ralf Poscher	Law	University of Freiburg	Europe
Michael Rosen	Government	Harvard	North America
Attorney Bana Shoughry-Badarne	Law	Hebrew University of Jerusalem	Asia
ICA 3 *Laws: Rigidity and Dynamics*			
Penelope Andrews	Law	New York Law School	North America
Partha Dasgupta	Economics	University of Cambridge	Europe
Michal Feldman	Computer Science	Tel Aviv University	Asia
Patrick Geary	History	IAS Princeton	North America
David Gross	Theoretical Physics	Kavli Institute for Theoretical Physics	North America
Wang Gungwu	History	National University of Singapore	Asia
Gil Kalai	Computer Science	Hebrew University of Jerusalem & Yale	Asia
Jean-Marie Lehn	Chemistry	University of Strasbourg	Europe
Atul N. Parikh	Biomedical Engineering	University of California, Davis	North America
Ernst Rank	Computation in Engineering	TUM Munich	Europe
Bo Rothstein	Political Science	University of Gothenburg	Europe
Michel Spiro	Physics	French National Center for Scientific Research	Europe
Geoffrey West	Physics	Santa Fe Institute	North America
Ada Yonath	Crystallography	Weizmann Institute of Science	Asia

Fig. 3.: Mentors and Speakers (ICA 2 and ICA 3).

Fellows and Supporting Institutions

ICA 1 *Time*			
Adriano De Cezaro	Mathematics	Federal University of Rio Grande	South America
Andre M. Cravo	Neuroscience	Federal University of ABC	South America
David Gange	History	University of Birmingham	Europe
Boris Roman Gibhardt	History of Art	Bielefeld University	Europe
Eduardo Almeida	Entomology	Universidade de São Paulo	South America
Eva von Contzen	English	Ruhr-University Bochum	Europe
Helder Nakaya	Molecular Biology	Universidade de São Paulo	South America
Kazuhisa Takeda	Latin America Studies	Waseda University	Asia
Yangyang Liu	Psychology	Nanjing University.	Asia
Marius Müller	Oceanography	Universidade de São Paulo	South America
Nikki Moore	Art History	Rice University	North America
Norihito Nakamichi	Molecular Biology	Nagoya University	Asia
Valtteri Arstila	Philosophy	University of Turku	Europe
ICA 2 *Human Dignity*			
Dr. Stephanie N. Arel	Theology	Boston University	North America
Guy E. Carmi	Lawyer	Lipa Meir & Co	Asia
Levi Cooper	Law	Tel Aviv University	Asia
Caterina Drigo	Law	University of Bologna	Europe
Dascha Düring	Philosophy	Utrecht University	Europe
Johannes Eichenhofer	Law	Universität Bielefeld	Europe
Connor M. Ewing	Politics	University of Texas at Austin	North America
Vanessa Hellmann	Law	Universität Bielefeld	Europe
Akemi Kamimura	Human Rights	University of São Paulo	South America
Sini Kangas	History	University of Tampere.	Europe
Michael Kolocek	Spatial Planning	TU Dortmund University	Europe
Sima Kramer	Law	Hebrew University of Jerusalem	Asia
Tamar Megiddo	Law	New York University	North America
Omer Michaelis	Philosophy	Tel Aviv University	Asia
Funlola Olojede	Theology	Stellenbosch University	Africa
Dr. Anita von Poser	Anthropology	Freie Universität Berlin	Europe
Dr. Caterina Preda	Political Science	University of Bucharest	Europe
Ron Roth	History	Ben-Gurion University Negrev	Asia
Talia Schwartz-Maor	Law	University of California, Berkeley	North America
Lauren Ware	Philosophy	University of Edinburgh	Europe
Emily Kidd White	Law	New York University	North America

Fig. 4.: Fellows and Supporting Institutions (ICA 1 and ICA 2).

Fellows and Supporting Institutions

ICA 3 *Laws: Rigidity and Dynamics*			
Vahid Aryadoust	Linguistics	National University of Singapore	Asia
V. V. Binoy	Neuroscience	National Institute of Advanced Studies (NIAS), Bangalore	Asia
Anupam Chattopadhyay	Computer Science	Nanyang Technological University	Asia
Willem Gravett	Law	University of Pretoria	Africa
Hanjo Hamann	Law	Max Planck Institute	Europe
Ulrich Heisserer	Engineering	DSM Materials Science Center, The Netherlands	Europe
Wai-Yip Ho	Sociology	The Education University of Hong Kong	Asia
Nkatha Kabira	Law	University of Nairobi, Kenya	Africa
Isabel Kusche	Sociology	Aarhus University, Denmark	Europe
Irina Kuznetsova	Human Geography	University of Birmingham, UK	Europe
Petra Liedl	Architecture	The University of Texas, Austin	Europe
David Maimon	Cyber-security	Georgia State University, USA	North America
Yoh Matsuo	Law	Nagoya University, Japan	Asia
Katrien Pype	Anthropology	University of Birmingham, UK	Europe
Tom Schonberg	Neuroscience	Tel Aviv University, Israel	Asia
Michael Smolkin,	Quantum Physics	The Hebrew University of Jerusalem	Asia
Carla A A Ventura	Nursing	University of São Paulo, Brazil	South America
Al Wilson	Philosophy	University of Birmingham, UK	Europe

Fig. 5.: Fellows and Supporting Institutions (ICA 3).

Carsten Dose   Dapeng Cai   Martin Grossmann   Regina Markus

Takao Kondo   Bernd Kortmann   Sami Pihlström   Eliezer Rabinovici

Till Roenneberg   Hisanori Shinohara

Fig. 6.: Orgnaisers of ICA 1.

Fernando Aith	Neka Menna Barreto	Tiago Bosisio Quental	Ruud Buijs
Massimo Canevacci	Sylvia Duarte Dantas	Carolina Escobar	Luiz Gylvan Meira Filho
Karl-Heinz Kohl	Fernando Iazzetta	Vera Lúcia Imperatriz Fonseca	Matthew Kleban

Fig. 7.: Mentors of ICA 1.

Eliezer Rabinovici

Takao Kondo	Hideyo Kunieda	Eduardo Monteiro	René Nome
Rodrigo Oliveira	Leopold Nosek	Suzana Pasternak	Renato Janine Ribeiro
Ana Lydia Sawaya	Paulo Saldiva	Laymert Garcia dos Santos	Hugo Segawa

Fig. 8.: Mentors of ICA 1.

Adriano De Cezaro

André M Cravo

David Gange

Boris Roman Gibhardt

Eduardo Almeida

Eva von Contzen

Helder Nakaya

Kazuhisa Takeda

Liu Yangyang

Marius Müuller

Nikki Moore

Norihito Nakamichi

Valtteri Arstila

Fig. 9.: Fellows of ICA 1.

Eliezer Rabinovici

| Ulrike Davy | Alon Harel | Hanna Lerner | Michal Linial |

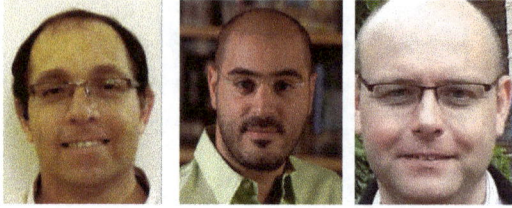

| Benny Porat | Nadiv Mordechay | Marc Schalenberg |

Fig. 10.: Orgnaisers of ICA 2.

Aleida Assmann

Admiral Ami Ayalon

Aharon Barak

Ben-Sasson

Meir Brezis

Bernadette Brooten

Marcus Duwell

Moshe Halbertal

Christine Hayes

Lynn A. Hunt

Michael Karayanni

Mordechai Kremnitzer

Gertrude Lübbe-Wolff

Raya Morag

Sari Nusseibeh

Ralf Poscher

Michael Rosen

Attorney Bana Shoughry-Badarne

Fig. 11.: Mentors of ICA 2.

Stephanie N. Arel   Guy E. Carmi   Caterina Drigo   Dascha Düring

Johannes Eichenhofer   Connor M. Ewing   Vanessa Hellmann   Akemi Kamimura

Sini Kangas   Michael Kolocek   Sima Kramer   Tamar Megiddo

Omer Michaelis   Funlola Olojede   Anita von Poser   Caterina Preda

Ron Roth   Talia Schwartz-Maor   Lauren Ware   Emily Kidd White

Fig. 12.: Fellows of ICA 2.

Lars Brink   Kwek Leong Chuan   Fiona de Londras   David Gange

Sue Gilligan   Michael J. Hannon   Phua Kok Khoo   Robin Mason

Eliezer Rabinovici   Jonathan Reinarz   Damien Walmsley

Fig. 13.: Organisers of ICA 3.

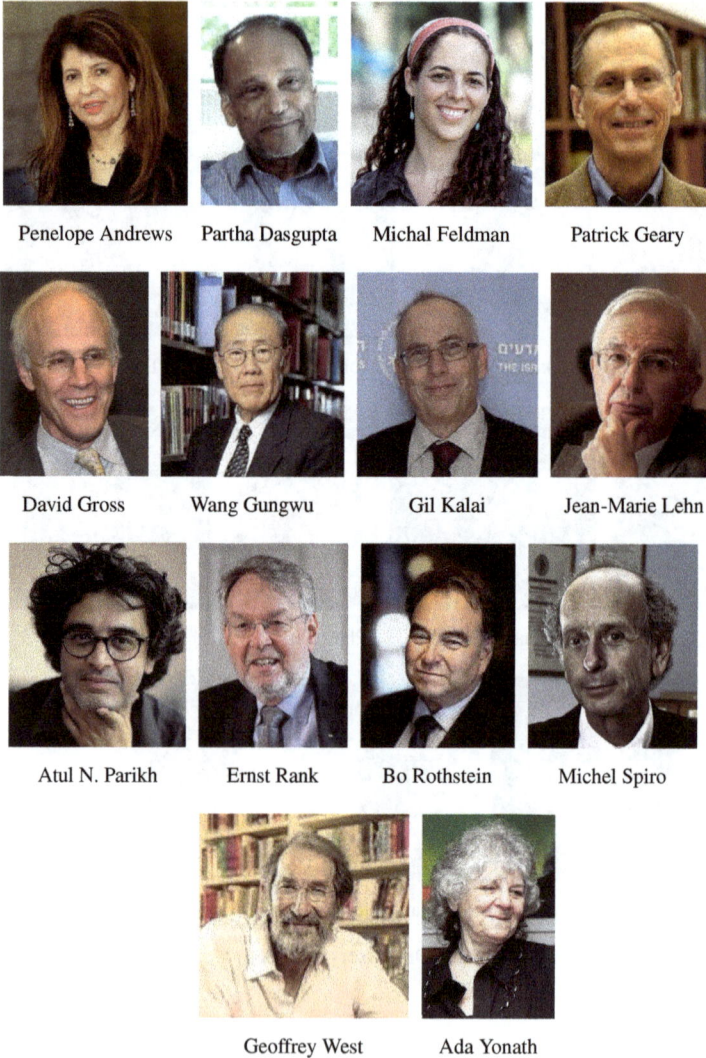

Penelope Andrews   Partha Dasgupta   Michal Feldman   Patrick Geary

David Gross   Wang Gungwu   Gil Kalai   Jean-Marie Lehn

Atul N. Parikh   Ernst Rank   Bo Rothstein   Michel Spiro

Geoffrey West   Ada Yonath

Fig. 14.: Mentors of ICA 3.

Vahid Aryadoust

V. V. Binoy

Anupam Chattopadhyay

Willem Gravett

Hanjo Hamann

Ulrich Heisserer

Wai-Yip Ho

Nkatha Kabira

Isabel Kusche

Irina Kuznetsova

Petra Liedl

David Maimon

Yoh Matsuo

Katrien Pype

Tom Schonberg

Michael Smolkin

Carla A A Ventura

Al Wilson

Fig. 15.: Fellows of ICA3.

Intercontinental Academia
São Paulo - April, 2015

schedule	Friday 17	Saturday 18	Sunday 19	Monday 20	Tuesday 21	Wednesday 22	Thursday 23	Friday 24	Saturday 25	Sunday 26	Monday 27	Tuesday 28	Wednesday 29	Thursday 30
8.30 am	Arrival	Scientific & cultural tour USP and Modernist São Paulo coordinated by Martin Grossmann, special participations by Paulo Saldiva and Hugo Segawa. Performance: Fernando Iazzetta	Scientific & cultural tour Peripheries -- Centralities (part 1) - with Ana Lydia Sawaya, Fernando Alth, Sylvia Dantas and Suzana Pasternak	Master class with José Goldemberg	Time off	Talk mediated by Vera Imperatriz Fonseca with Tiago Quental and Luiz Gylvan Meira Filho	Resting day with social-cultural activities	Time off	Talk with Leopold Nosek	Raising questions	Time off	Time off	Time off	Check out
9.00 am								Breakfast at USP						
9.30 am					Talk with Sami Pihlström									
10.00 am				Break		Break		Workshop with brazilian Minister of Education, Renato Janine Ribeiro	Break	Break	Designing a MOOC on time	Designing and Proposals	Closing report	
10.30 am				Talk with Matthew Kleban	Talk with Carolina Escobar	Talk with Karl-Heinz Kohl			Plenary	Raising questions				
11.00 am													Talk with Scientific Committe	
11.30 am														
12.00 pm				Talk about our MOOC	Participants' presentation	Participants' presentation			Participants' presentation	Participants' presentation				
12.15 pm		Lunch	Lunch at Mocoto with Rodrigo de Oliveira								Lunch at the President's Office	Lunch	Lunch	
1.00 pm				Lunch	Lunch	Lunch		Lunch	Lunch	Lunch				
1.45 pm					Participants' presentation	Participants' presentation			Participants' presentation	Participants' presentation				
2.00 pm			Scientific & cultural tour Peripheries -- Centralities (part 2) - with Ana Lydia Sawaya, Fernando Alth, Sylvia Dantas and Suzana Pasternak	Talk with Laymert Garcia dos Santos		Wrap-up of the previous talks		Time off	Talk with Till Roenneberg about his experience with Coursera	Talk with Regina P. Markus	Designing and Proposals	Designing and Proposals		
2.30 pm					Talk with Ruud Buijs						Talk with Massimo Canevacci			
3.00 pm				Break										
3.30 pm					Break	Break			Break	Break				
4.00 pm				Talk with Renê Nome	Talk with Hideyo Kunieda	Talk with Course Success Team of Coursera		The Future of the Universities with Marco A. Zago, Carlos Vogt, Naomar de Almeida Filho, Luiz Bevilacqua, João Heraldo de Lima Capella, Deborah Helena B. Nader and Marcelo Knobel, moderator Sabine Righetti	Raising questions	Plenary	Designing and Proposals	Designing and Proposals	Time off	
4.30 pm														
5.00 pm									Talk with Vera Lucia Imperatriz Fonseca					
5.30 pm				Break	Break	Break			Time off	Break	Break			
6.00 pm	Transfer to USP's School of Medicine	Time off	Time off	Talk with Eliezer Rabinovici	Talk with Till Roenneberg	Talk with Takao Kondo		Report by Marcelo Knobel		Plenary	Designing and Proposals			
6.30 pm														
7.00 pm	Opening with Renato J. Ribeiro, José E. Krieger, Hernan Chalmovich, Cai Dapeng, Carsten Dose and Martin Grossmann at USP's School of Medicine													
7.30 pm														
8.00 pm		Dinner at the hotel	Dinner at the hotel	Dinner at the hotel	Dinner at the hotel	Dinner at the hotel	Dinner at the hotel	Dinner at the hotel	Dinner at the hotel	Dinner at the hotel	Dinner at the hotel	Dinner at the hotel	Closing dinner with Neka Menna Barreto	
8.30 pm														
9.00 pm														

INSTITUTE FOR ADVANCED RESEARCH, NAGOYA UNIVERSITY

UBIAS Intercontinental Academia Nagoya Workshop
MARCH 6-18, 2016

INTERCONTINENTAL ACADEMIA

WEEK 1 / MARCH 6-12

SUNDAY, MARCH 6	MONDAY, MARCH 7	TUESDAY, MARCH 8	WEDNESDAY, MARCH 9	THURSDAY, MARCH 10	FRIDAY, MARCH 11	SATURDAY, MARCH 12
9:30 a.m.–12:00 p.m. **Script Workshop by the Participants 1** (participants only) Seminar Room, Sci. Building B	**Opening Ceremony** (Chair: Hitoshi Sakakibara, Nagoya IAR) 9:30 a.m.–9:40 a.m. **Welcome Remarks** by Seiichi Matsuo, President, Nagoya University **Group Photograph** 9:50 a.m.–10:20 a.m. **Lecture Academic Research at Nagoya University** by Hideyo Kunieda, Trustee and Vice President, Nagoya University 10:20 a.m.–10:40 a.m. **Welcome Remarks** by Martin Grossmann, Director, IEA-USP, and Carsten Dose, Managing Director, FRIAS 10:40 a.m.–11:00 a.m. **Nagoya IAR's Activities** by Hisanori Shinohara, Director, Nagoya IAR 11:10 a.m.–12:00 p.m. **Campus Tour: "Nobel Road"** with Hisanori Shinohara, Nagoya IAR	**Biology Workshop** (Chair: Hitoshi Sakakibara, Nagoya IAR) 9:00 a.m.–9:10 a.m. **Opening Remarks** by Takao Kondo, Nagoya University 9:10 a.m.–10:00 a.m. **Lecture** by Kenichi Honma, Hokkaido University (Chair: Till Roenneberg, Ludwig Maximilians University) 10:10 a.m.–11:00 a.m. **Lecture** by Hideharu Numata, Kyoto University (Chair: Takashi Yoshimura, Nagoya University) 11:20 a.m.–12:00 p.m. **Laboratory Tour** with Norihito Nakamichi Institute of Transformative Bio-Molecules (ITbM) Hall	**Physics Workshop** 8:45 a.m.–9:10 a.m. **Opening Remarks and Overview** by Naoshi Sugiyama, Nagoya IAR 09:10 a.m.–10:00 a.m. **Lecture** by Takanori Sasaki, Kyoto University (Chair: Hideyo Kunieda, Nagoya University) 10:10 a.m.–11:00 a.m. **Lecture** by Masao Takamoto, RIKEN (Chair: Hideyo Kunieda, Nagoya University) 11:00 p.m.–11:50 a.m. **Lecture** by Tadashi Takayanagi, Kyoto University (Chair: Eliezer Rabinovici, Hebrew University in Jerusalem) Environmental Studies Hall	**Humanities/Social Sciences Workshop** (Chair: Hajime Wada, Nagoya IAR) **Part I: Time in Humanities and Social Sciences** 9:00 a.m.–9:10 a.m. **Opening Remarks** by Takaho Ando, Chubu University 9:10 a.m.–10:00 a.m. **Lecture** by Yoshiyuki Suto, Nagoya University 10:10 a.m.–11:00 a.m. **Lecture** by Yasuhira Kanayama, Nagoya University 11:10 a.m.–12:00 a.m. **Lecture** by Takehiro Ohya, Keio University Environmental Studies Hall	9:30 a.m.–11:00 a.m. **Panel Discussion by the Participants Interdisciplinarity: Benefits and Challenges of the Intercontinental Academia** *Moderator: Eva von Contzen, FRIAS* 11:10 a.m.–12:00 p.m. **Meeting with the Senior Committee** Environmental Studies Hall	9:00 a.m. **Explore the City of Nagoya** (Bus leaves at 9:00 a.m. from the Toyota Auditorium, Nagoya University) Toyota Commemorative Museum of Industry and Technology; Tokugawa Art Museum
	12:00 p.m.–1:00 p.m. **Lunch** Seminar Room, 1F, Sakata and Hirata Hall	12:00 p.m.–1:00 p.m. **Lunch** Restaurant Hanaoki	12:00 p.m.–1:00 p.m. **Lunch Talk by Eliezer Rabinovici, Hebrew University in Jerusalem** Restaurant Hanaoki	12:00 p.m.–1:00 p.m. **Lunch Talk by Sami Pihlström, University of Helsinki** Restaurant Hanaoki	12:00 p.m.–1:00 p.m. **Lunch** Restaurant Hanaoki	12:00 p.m.–1:00 p.m. **Lunch** Kouyouen
(Bus leaves at 6:00 p.m. from the Toyota Auditorium, Nagoya University/ANA Hotel) 6:30 p.m.–8:30 p.m. **Reunion Reception** Kisoji Yagoto	1:30 p.m.–2:30 p.m. *Chair: Hisanori Shinohara, Nagoya IAR* **Master Class with Nobel Laureate Toshihide Maskawa, Nagoya University** *(Moderator: Naoshi Sugiyama, Nagoya University/ U=E, S)* 3:00 p.m.–4:00 p.m. **Master Class with Nobel Laureate Ryoji Noyori, Nagoya University (E>L S)** 4:20 p.m.–4:50 p.m. **Premiere A Documentary on ICA São Paolo Workshop** 4:50 p.m.–5:50 p.m. **Introduction of ICA and Participants' Research** 6:30 p.m.–7:00 p.m. **Keynote Speech Higher Education and Academic Research in Japan** by Michinari Hamaguchi, President, JST Sakata and Hirata Hall 7:10 p.m.–8:30 p.m. **Welcome Reception** Restaurant Hanaoki	1:20 p.m.–2:10 p.m. **Lecture** by Hideo Iwasaki, Waseda University (Chair: Martin Grossmann, University of São Paulo) 2:10 p.m.–3:00 p.m. **Lecture** by Kazuhiko Kume, Nagoya City University (Chair: Hitoshi Sakakibara, Nagoya IAR) 3:10 p.m.–3:35 p.m. **Challenging the "Box"** 3:35 p.m.–4:10 p.m. **Discussion with Speakers** Environmental Studies Hall	1:20 p.m.–2:10 p.m. **Lecture** by Satoshi Nozawa, Dokkyo University (Chair: Eliezer Rabinovici, Hebrew University in Jerusalem) 2:10 p.m.–2:35 p.m. **Challenging the "Box"** 2:35 p.m.–3:10 p.m. **Discussion with Speakers** Environmental Studies Hall	**Part II: Oriental Time** 1:20 p.m.–1:30 p.m. **Opening Remarks by Takaho Ando, Chubu University** 1:30 p.m.–2:30 p.m. **Lecture** by Chun-chieh Huang, National Taiwan University 2:40 p.m.–3:30 p.m. **Lecture** by Kirill O. Thompson, National Taiwan University 3:30 p.m.–3:55 p.m. **Challenging the "Box"** 3:55 p.m.–4:30 p.m. **Discussion with Speakers** Environmental Studies Hall	*Chair: Carsten Dose, FRIAS* 1:00 p.m.–2:30 p.m. **Keynote Speech The Development of Institutes for Advanced Study and their Role in the Contemporary University** by Peter Goddard, Former Director, Institute for Advanced Study 2:50 p.m.–4:50 p.m. **Panel Discussion The Future of UBIAS** by Peter Goddard (IAS), Till Roenneberg (Munich), Kirill O. Thompson (Taiwan), Sari Kivistö (Helsinki), Jonathan Reharz (Birmingham), Sue Gilligan (Birmingham), Eliezer Rabinovici (Jerusalem), Martin Grossmann (IEA-USP), Hideaki Miyajima (WIAS), and Hisanori Shinohara (Nagoya IAR) Environmental Studies Hall	(Bus returns to Kanayama Station after lunch at around 2:00 p.m.) **Free Time**
		4:30 p.m.–6:30 p.m. **Consolidation Workshop I** (participants only) Seminar Room, Sci. Building B	3:30 p.m.–6:30 p.m. **Consolidation Workshop II** (participants only) Seminar Room, Sci. Building B	4:50 p.m.–6:30 p.m. **Consolidation Workshop III** (participants only) Seminar Room, Sci. Building B	5:30 p.m.–6:10 p.m. **Zen Meditation (Zazen)** Yagoto Koushoji Temple 7:00 p.m.–8:30 p.m. **Banquet** Yagoto Koushoji Temple	

INSTITUTE FOR ADVANCED RESEARCH, NAGOYA UNIVERSITY
UBIAS Intercontinental Academia Nagoya Workshop
MARCH 6-18, 2016

INTERCONTINENTAL ACADEMIA — UBIAS

WEEK 2 / MARCH 13-19	SUNDAY, MARCH 13	MONDAY, MARCH 14	TUESDAY, MARCH 15	WEDNESDAY, MARCH 16	THURSDAY, MARCH 17	FRIDAY, MARCH 18	SATURDAY, MARCH 19
Morning	Hotel checking out Nagoya to Tokyo by the Shinkansen Bullet Train (100 minutes) Explore the City of Tokyo Free Time	**Waseda Workshop In Search of Interdisciplinary Dialogue** organized by WIAS 早稲田大学高等研究所 **Opening Remarks** by Hideaki Miyajima, Director, WIAS 9:30 a.m.–9:45 a.m. **Keynote Lecture** by Till Roenneberg, Ludwig-Maximilians University 9:45 a.m.–10:25 a.m. **WIAS Lectures on TIME from an Interdisciplinary Perspective** by Yu Tahara, Ryota Akiyoshi and Masashi Abe, WIAS 10:45 a.m.–12:15 p.m. *Moderator: Kanami Orihara, WIAS* **Q&A Session** CTLT Classroom 2-3, 2nd floor, Bldg. 3, Waseda University	**Arts Workshop** 9:00 a.m.–9:10 a.m. **Opening Remarks** by Takaho Ando, Chubu University and Nagoya University (J+E, C) 9:10 a.m.–9:30 a.m. *Moderator: Martin Grossmann, University of São Paulo (E+J, C)* 10:00 a.m.–12:00 p.m. **Lecture** by Satoru Kitagou and Akitoshi Edagawa, Tokyo University of the Arts (J+E, C) Environmental Studies Hall	**Coursera Session** 9:00 a.m.–10:00 a.m. IAR Hall 10:20 a.m.–12:00 p.m. **Workshop Present and Causality** (participants only) Seminar Room, Sci. Building B	9:30 a.m.–12:00 p.m. **Workshop Relativity and Individuality** (participants only) Seminar Room, Sci. Building B	9:30 a.m.–12:00 p.m. **Workshop by the Participants Final Presentation** Sakata and Hirata Hall	
Lunch		12:30 p.m.–1:30 p.m. **Lunch** Good Morning Cafe	12:00 p.m.–1:00 p.m. **Lunch** Restaurant Hanaoki	12:00 p.m.–1:00 p.m. **Lunch** IAR Lounge, Sci. Building B	12:00 p.m.–1:00 p.m. **Lunch** IAR Lounge, Sci. Building B	12:00 p.m.–1:00 p.m. **Lunch** Restaurant Hanaoki	
Afternoon	Free Time	1:30 p.m.–2:00 p.m. **Campus Tour** (optional) 2:30 p.m.–4:45 p.m. **Discussion and Presentation** **Comments and Closing Remarks** by Hatsue Shinohara, Associate Dean of Research Promotion Division, Waseda University *Moderator: Mariko Kellom, WIAS* CTLT Classroom 2-3, 2nd floor, Bldg. 3, Waseda University **Tokyo to Nagoya by the Shinkansen Bullet Train** (100 minutes)	2:00 p.m.–5:00 p.m. **Japanese Tea Ceremony (Chado)** **Opening Remarks** by Kuong Teilee, Nagoya University **Closing Remarks** by Hajime Wada, Nagoya University (J+E, C) CALE Hall Tearoom	1:30 p.m.–3:30 p.m. **Workshop Measurement and Value** (participants only) 4:00 p.m.–6:00 p.m. **Workshop Rhythms and History** (participants only) Seminar Room, Sci. Building B	1:30 p.m.–3:30 p.m. **Workshop Illusions, Endings, and the Future** (participants only) 4:00 p.m.–6:00 p.m. **Workshop Traces and Representations** (participants only) Seminar Room, Sci. Building B	1:30 p.m.–3:00 p.m. **Meeting of the Participants** 3:30 p.m.–5:30 p.m. **Final Meeting with the UBIAS Intercontinental Academia Senior Committee** 5:30 p.m.–6:30 p.m. **Final Meeting of the Participants** Sakata and Hirata Hall	
Evening						6:30 p.m. **Closing Dinner** chez Jroud	
	Hotel RIHGA Royal Hotel Tokyo						

GREEN WORKSHOP

Intercontinental Academia Nagoya Workshop will be organized as a green workshop. This implies that we will try to keep the ecological impact of the workshop as low as possible by avoiding unnecessary trash, using recyclable materials and public transportation, etc. Please, have a hand in this endeavor and take into account environment friendly options in planning your participation.

Israel Institute for Advanced Studies (IIAS)
Zentrum für interdisziplinäre Forschung (ZiF)

Intercontinental Academia on Human Dignity

Phase 1

March 6-18, 2016

All lectures will take place at the IIAS, The Hebrew University of Jerusalem, Edmond J. Safra Campus, Givat Ram (Room 128)

Organizers:

Michal Linial (Israel Institute for Advanced Studies)
Ulrike Davy (Bielefeld University)

Program

Sunday, 6 March

08:30 - 09:00	**Registration** IIAS Lobby
09:00 - 10:30	**Michal Linial** (Israel Institute for Advanced Studies) Opening Reception (Fellows Lounge)
10:30 - 12:30	**Marcus Duwell** (Utrecht University) Introductory Round (Room 128)
12:30 - 14:00	**Lunch** The Belgium House (on campus)
14:00 - 17:00	**Marcus Duwell** (Utrecht University) Concepts and Discourses around Human Dignity (Room 128)
17:00 - 19:00	**A sightseeing tour of Jerusalem**
19:00 - 20:30	**Opening Dinner at Eucalyptus restaurant** (With Michal, Ulrike and Marcus)

Monday, 7 March

09:00 - 12:30	**Marcus Duwell** (Utrecht University) A Kantian Account of Human Dignity (Room 128)
12:30 - 14:00	**Lunch** (IIAS, Room 115)
14:00 - 17:00	**Marcus Duwell** (Utrecht University) Human Dignity as the Foundation of Human Rights (Room 128)

2

Eliezer Rabinovici

17:00 - 18:45	**Optional Social Event** IIAS

Tuesday, 8 March

09:00 - 12:30	**Marcus Duwell** (Utrecht University) Human Dignity and Future Generations (Room 128)
12:30 - 14:00	**Lunch** (IIAS, Room 115)
14:00 - 17:00	**Marcus Duwell** (Utrecht University) Human Dignity: China and the West (Room 128)
17:00 - 17:35	**Light Dinner at the IIAS**
17:35 - 20:00	**Tour of the Israel Museum**

Wednesday, 9 March

10:00 - 10:25	Transportation from YMCA to Givat Ram
10:30 - 12:00	**Tour of the Edmond J. Safra Campus, Givat Ram** (With Prof. Jeff Camhi)
12:00 - 14:30	**Bernadette Brooten** (Brandeis University) Introductory Round (with Benny Porat) (Room 128)
14:30 - 15:15	**Lunch** IIAS Fellows Lounge
15:15 - 18:30	**Bernadette Brooten** (Brandeis University) The Roman Catholic Church: From Slavery to Support for Workers' Dignity (Room 128)

18:30 - 19:30	**Raya Morag (Hebrew University)** Representations of Human Rights in Israeli Cinematography during the Intifada (Room 128)
19:45 - 20:45	Transportation from Givat Ram to the YMCA

Thursday, 10 March

08:30 - 09:00	Transportation from YMCA to the Suprem Court of Israel
09:00 - 10:15	**A visit to the Supreme Court of Israel**
10:15 - 10:35	Transportation from the Supreme Court to Givat Ram
10:30 - 14:30	**Bernadette Brooten** (Brandeis University) Slavery, Religion, and Women's Dignity (Room 128)
14:30 - 15:15	**Lunch** The Belgium House (on campus)
15:15 - 18:30	**Bernadette Brooten** (Brandeis University) "The Dignity of Created Beings" ('kvod habriot') as an Emerging Legal Concept in Jewish Law (Room 128)
18:30 - 19:00	Transportation to Tcenim Restaurant
19:00 - 20:30	**Dinner at Te'enim restaurant** With Prof. Brooten

Friday, 11 March

08:30 - 09:00 Transportation from YMCA to Yad Vashem

09:00 - 12:00 **A visit to Yad Vashem**

12:00 - 12:30 Transportation from Yad Vashem to Ima Restaurant

12:30 - 13:30 **Lunch at Ima restaurant**

13:30 - 20:00 Free Time

Saturday, 12 March

08:00 - 15:25 **A one-day excursion to Massada (and lunch at Ein Gedi)** Bus will depart at 8:00.(Tour paid by participants, NIS 100)

Sunday, 13 March

08:00 - 08:25 Transportation from YMCA to Givat Ram

08:30 - 13:15 **Christine Hayes** (Yale University), **Moshe Halbertal** (The Hebrew University of Jerusalem) Human Dignity in Jewish Sources and Jewish Thought (Room 128)

13:15 - 14:00 **Lunch** The Belgium House (on campus)

14:00 - 16:30 **a visit to the The Residence of the President of Israel** and a talk with Mr. David Saranga

16:30 - 18:00 **Christine Hayes** (Yale University) (Room 128)

18:00 - 18:40 **Light Dinner at the IIAS**

18:40 - 20:30 **A talk with Prof. Sari Nusseibeh, Professor of Philosophy & former President of Al-Quds University** (Room 128)

20:45 - 21:45 Transportation from Givat Ram to YMCA

Monday, 14 March

08:30 - 09:00 Transportation from YMCA to Givat Ram

09:00 - 12:30 **Aharon Barak** (Interdisciplinary Center (IDC), Herzliya) Human Dignity as a Constitutional Value (Room 128)

12:30 - 14:00 **Lunch** (IIAS, Room 115)

14:00 - 17:00 **Aharon Barak** (Interdisciplinary Center (IDC), Herzliya) Human Dignity as a Constitutional Right (Room 128)

17:00 - 19:00 **Keynote speaker** TBD

Tuesday, 15 March

09:00 - 10:00 Free

10:00 - 12:30	**Michael Rosen** (Harvard University) Dignity, It's History and Meaning: Part 1 (Room 128)
12:30 - 14:00	**Lunch** At the IIAS
14:00 - 16:30	**Michael Rosen** (Harvard University) Dignity, It's History and Meaning: Part 2 (Room 128)
16:30 - 19:00	**Tour of Nachlaot and Mahane Yehuda Market in Jerusalem** Tourguide: Guy Noyman Meet at the IIAS Lobby
19:00 - 20:30	**Dinner in Trattoria Haba restaurant**
20:30 - 21:00	Transportation from Trattoria Haba restaurant to YMCA

Wednesday, 16 March

08:30 - 09:00	Transportation from YMCA to Givat Ram
09:00 - 10:30	**Menahem Ben-Sasson** (The Hebrew University of Jerusalem) The Meeting Point between The Constitutional Charter of Rights, the Article on Override and Human Dignity – A Testimony (Room 128)
10:30 - 12:30	**Mordechai Kremnitzer** (The Hebrew University of Jerusalem) Human Dignity and National Security: Introduction, and a talk with Admiral Ami Ayalon (former head of the Shin Bet)
12:30 - 14:00	**Lunch** IIAS Fellows Lounge
14:00 - 15:30	**Mordechai Kremnitzer** (The Hebrew University of Jerusalem) Balancing the (evolving) Right to Dignity with National Security in Israel (Room 128)
15:30 - 17:00	**Mordechai Kremnitzer** (The Hebrew University of Jerusalem) On the Dignity of Groups and the Dignity of Individuals: The Status of Palestinian–Arabs under Israel's Constitutional Definition as a "Jewish and Democratic" State - Prof. Michael Karayanni
17:30 - 18:00	Transportation from Givat Ram to Hachatzer Restaurant
18:00 - 19:30	**Dinner at Hachtzer Restaurant** (with Prof. Kremnitzer)

Thursday, 17 March

08:30 - 09:00	Transportation from YMCA to Givat Ram
09:00 - 10:30	**Mordechai Kremnitzer** (The Hebrew University of Jerusalem) HCJ 5100/94 Public Committee Against Torture in Israel et al. v. The State of Israel et al. Judgment (1999) - With Adv. Banna Shugri Badarne
10:30 - 12:30	**Mordechai Kremnitzer** (The Hebrew University of Jerusalem) Challenges to Human Dignity in Health Care - with Prof. Mayer Brezis

PAGE 2

ZiF
Zentrum für interdisziplinäre Forschung
Center for Interdisciplinary Research
Universität Bielefeld

PROGRAMME

ZiF SUMMER SCHOOL

Intercontinental Academia on Human Dignity

Conveners: Prof. Ulrike Davy (Bielefeld, GER), Prof. Michal Linial (Jerusalem, ISR)

1 – 11 August 2016

INTERCONTINENTAL ACADEMIA

ZiF
Zentrum für interdisziplinäre Forschung
Center for Interdisciplinary Research
Universität Bielefeld

MONDAY 1st AUGUST 2016

MASTERCLASS G:
THE CONSTITUTIONAL CLAUSE ON RESPECTING HUMAN DIGNITY (ARTICLE 1 GERMAN GRUNDGESETZ)

9:00 – 9:30	Welcome addresses: Martin Egelhaaf (Vice-Rector of Bielefeld University), Marc Schalenberg (ZiF), Michal Linial (IIAS), Ulrike Davy (Bielefeld University)
9:30 – 11:00	Gertrude Lübbe-Wolff: Overview on the meaning of the German dignity clause I
11:00 – 11:30	*Coffee break*
11:30 – 12:30	Gertrude Lübbe-Wolff: Overview on the meaning of the German dignity clause II
12:30 – 14:00	*Lunch break*
14:00 – 15:00	Human dignity in cross-country perspective: Emily Kidd White (Canada)
15:00 – 16:30	Privatisation of prisons and human dignity: Sima Kramer (Israel), Gertrude Lübbe-Wolff
16:30 – 17:00	*Coffee break*
17:00 – 18:30	Fellows' input: Prisoners' dignity: Vanessa Hellmann (Germany), Akemi Kamimura (Brazil)
18:30 – 19:30	What is good to know about ZiF, Bielefeld and Germany
19:30	*Barbecue at ZiF*

TUESDAY, 2nd AUGUST 2016

9:30 – 11:00	Ralf Poscher: The German dignity clause: Cases, arguments, European context I
11:00 – 11:30	*Coffee break*
11:30 – 12:30	Ralf Poscher: The German dignity clause: Cases, arguments, European context II
12:30 – 14:00	*Lunch break*
14:00 – 15:00	Ralf Poscher: The German dignity clause: Cases, arguments, European context III
15:00 – 16:30	Fellows' input: Where could human dignity do wrong: Guy Carmi (Israel)
16:30 – 17:00	*Coffee break*
17:00 – 18:30	Fellows' input: Human dignity and privacy: Johannes Eichenhofer (Germany)
18:30 – 19:30	*Dinner*
20:00	Walk through Bielefeld

WEDNESDAY, 3rd AUGUST 2016

9:30 – 11:00	Ulrike Davy: Refugee crisis 2015/16 and human dignity I
11:00 – 11:30	*Coffee break*
11:30 – 12:30	Ulrike Davy: Refugee crisis 2015/16 and human dignity II
12:30 – 14:00	Lunch break
14:00 – 15:00	Ulrike Davy: Refugee crisis 2015/16 and human dignity III
15:00 – 16:30	Fellows' input: Social rights of immigrants: Caterina Drigo (Italy); The right to housing: Michael Kolocek (Germany)
16:30 – 17:00	*Coffee break*
17:00 – 18:30	Video artist Anna Konik: In the same city, under the same sky – interviews with immigrants
18:30 – 19:30	*Dinner*
20:00	Fellow room: Afterthoughts with Anna Konik

SATURDAY, 6TH AUGUST 2016

9:30 – 11:00   **Lynn A. Hunt**: The powerful paradox of universalism I

11:00 – 11:30   *Coffee break*

11:30 – 12:30   **Lynn A. Hunt**: The powerful paradox of universalism II

12:30 – 13:30   *Lunch break*

14:00   Time out (trip to Berlin, self pay)

SUNDAY, 7TH AUGUST 2016

Time out, no service at ZiF

MONDAY, 8TH AUGUST 2016

Breakfast available (8:00 – 9.00 a.m.)

Time out, no service at ZiF

19:00   Barbecue together with Summer School "Randomness in Physics & Mathematics"

TUESDAY, 9TH AUGUST 2016

MASTERCLASS I:
RECOGNIZING HUMAN DIGNITY BEFORE ITS INVENTION AND AFTER ITS DENIAL

9:30 – 11:00   **Aleida Assmann**: Empathy and its limits

11:00 – 11:30   *Coffee break*

11:30 – 12:30   **Aleida Assmann**: Liminal anthropology in Shakespeare's tragedies

12:30 – 14:00   *Lunch break*

14:00 – 16:30   **Aleida Assmann**: Sameness, similarity and dignity (task groups)

16:30 – 17:00   *Coffee break*

17:00 – 18:30   **Fellows' Input**: Trauma and human dignity: **Stefanie Arel** (United States of America)

18:30 – 19:30   *Dinner*

20:00   Fellow room: Afterthoughts with **Aleida Assmann**

THURSDAY, 4TH AUGUST 2016

MASTERCLASS H:
HUMAN DIGNITY AS THE CORE OF HUMAN RIGHTS

9:30 – 11:00   **Lynn A. Hunt**: Why the history of human dignity / human rights matters I

11:00 – 11:30   *Coffee break*

11:30 – 12:30   **Lynn A. Hunt**: Why the history of human dignity / human rights matters II

12:30 – 14:00   *Lunch break*

14:00 – 16:30   **Lynn A. Hunt**: Human dignity and human rights: What does equality mean? (autonomy, empathy, respect) (task groups)

16:30 – 17:00   *Coffee break*

17:00 – 18:30   **Fellows' Input**: Human dignity and the emotions of fear and suffering: **Lauren Ware** (United Kingdom); Freedom of religion in art: The case of Napoleon and the Jews: **Levi Cooper** (Israel)

18:30 – 19:30   *Dinner*

20:00   Movie night : *Footnote*

FRIDAY, 5TH AUGUST 2016

9:30 – 11:00   **Lynn A. Hunt**: Frames of personhood and "secularism" I

11:00 – 11:30   *Coffee break*

11:30 – 12:30   **Lynn A. Hunt**: Frames of personhood and "secularism" II

12:30 – 14:00   *Lunch break*

14:00 – 16:30   **Lynn A. Hunt**: How does a longer-term view change our understanding of current dilemmas? (task groups)

16:30 – 17:00   *Coffee break*

17:00 – 18:30   **Fellows' Input**: Non-combatants in medieval Christian holy war: **Sini Kangas** (Finland)

18:30 – 19:30   *Dinner*

WEDNESDAY, 10TH AUGUST 2016

9:30 – 11:00	**Aleida Assmann:** Cultures of remembrance – the 'German model'
11:00 – 11:30	*Coffee break*
11:30 – 12:30	**Aleida Assmann:** Counter monuments
12:30 – 14:00	*Lunch break*
14:00 – 16:30	**Aleida Assmann:** The spectral turn (task groups) I
16:30 – 17:00	*Coffee break*
17:00 – 18:30	**Aleida Assmann:** The spectral turn (task groups) II
18:30 – 19:30	*Dinner*
20:00	Walk through Bielefeld

THURSDAY, 11TH AUGUST 2016

9:30 – 11:00	**Aleida Assmann:** Civilizing societies
11:00 – 11:30	*Coffee break*
11:30 – 12:30	**Aleida Assmann:** Human rights and human responsibilities
12:30 – 14:00	*Lunch break*
14:00 – 18:00	**Fellows' Session**
18:30	Fellow Room: *Farewell Dinner*

NANYANG TECHNOLOGICAL UNIVERSITY

Institute of Advanced Studies

INTERCONTINENTAL ACADEMIA

UNIVERSITY OF BIRMINGHAM

THE INSTITUTE OF ADVANCED STUDIES

Intercontinental Academia Singapore March 19-27 2018

Orchid Room, NTU Campus Clubhouse, 50 Nanyang View, Singapore 639667

Sunday 18 March	18:30 Welcome Reception @ Cosmo		
Monday 19 March	**Morning**	**Afternoon**	**Evening**
Fellows All Fellows Mentors Professor Michal Feldman Professor David Gross Professor Patrick Geary Professor Gil Kalai Professor Michel Spiro Directors Professor Lars Brink Professor Kwek Leong Chuan Professor Michael J Hannon Professor Phua Kok Khoo Professor Eliezer Rabinovici Birmingham Organising Group Professor Damien Walmsley (University of Birmingham) Guests Professor Nargiza Amirova (IAR Nagoya University)	**Welcome** **09:00-09:30** Phua Kok Khoo Michael J Hannon Eliezer Rabinovici Kwek Leong Chuan Sue Gilligan Briefing on: Mentor talks responses 1:1s with Mentors Thematic Panels Projects and outcomes Presentation to Brazil **09:30-12:00** Fellows introductions to each other Mind-mapping and planning **Includes 10:30 Refreshments break** **12:00 Morning wrap up**	**Mentor talks and responses** **Chair - Eliezer Rabinovici** 13:30-15:30 *Michal Feldman* *Algorithmic Game Theory: the Crossroad of Computer Science, Game Theory, and Economics* Response and discussion *Michel Spiro + two fellows* 15:30 Refreshments 15:45-17:45 *Patrick Geary* *Laws of History and Laws in History* Response and discussion *David Gross + two fellows*	**Dinner** **Catering @ Orchid Room** **18:30** **20:00-21:00 Evening discussion** All fellows Michal Feldman Patrick Geary

Tuesday 20 March	Morning	
	Fellows' 6 minute Presentations	10:40-11:00 Refreshments
Fellows All Fellows	**09:00-09:30** Dr Vahid Aryadoust *Why aren't there any "Scientific Laws" in Applied Linguistics?"* Dr VV Binoy *Determinants of Social Decision-making: a Tale of Two Vertebrates* Dr Anupam Chattopadhyay *Laws of Digital Evolution*	**11:00-11:30** Dr Petra Liedl *Laws and Sustainability- How do Laws Support Transformation Processes?* Dr Kenneth Lim *Intuitions and Learning, and (very) early musings on Intuitions and Laws* Dr David Maimon *Regulating Computer Users' Online Behaviors: Online and Offline Architectures of Control*
Mentors Professor Penny Andrews Professor Michal Feldman Professor Patrick Geary Professor David Gross Professor Gil Kalai Professor Atul Parikh Professor Michel Spiro	**09:30-10:00** Dr Willem Gravett *The Legal System's Uneasy Relationship with Social Science: Can Rigidity Make Way for a Dynamic?* Dr Hanjo Hamann *Why European Lawyers Call Their Trade a "Science", and What That Tells Us About Their Dynamical Systems Perspective* Dr Uli Heisserer *Industrial Research in Materials Science: Doing Meaningful Things where Laws of Nature (science), Laws of Society (legislation, ethics) and Market Laws Interplay*	**11:30-12:00** Professor Yoh Matsuo *Two Concepts of Law: Legal and Scientific* Dr Katrien Pype *Mobile Phones: the Matter of Communication* Dr Tom Schonberg *New Paradigms, Open Science and Neuroethics of "mind-reading"*
Directors Professor Lars Brink Professor Kwek Leong Chuan Professor Michael J Hannon Professor Phua Kok Khoo Professor Eliezer Rabinovici Birmingham Organising Group Professor Damien Walmsley (University of Birmingham) Guests Professor Nargiza Amirova (IAR Nagoya University)	**10:00-10:40** Dr Wai-Yip Ho *Rigidity and Dynamics: Considering Islamic Law* Dr Nkatha Kabira *Commissions as Technologies of Law and Governance* Dr Isabel Kusche *Legal Rules: Content versus Meaning* Dr Irina Kuznetsova *Law, Mobility and Bodies: Developing an Interdisciplinary Theoretical and Methodological Approach to Studying the Outcomes of Displacement*	**12:00-12:30** Dr Michal Smolkin *Law through the Lens of Physics* Dr Carla Aparecida Arena Ventura *International Law and International Human Rights Law: Exploring Interrelationships, Rigidity and Dynamics in the areas of Health and Mental Health* Dr Al Wilson *Laws of Nature: Unity and Variety*

Tuesday 20 March (cont.)	Afternoon	Evening
Fellows All Fellows **Mentors** Professor Penny Andrews Professor Michal Feldman Professor Patrick Geary Professor David Gross Professor Gil Kalai Professor Atul Parikh Professor Michel Spiro **Directors** Professor Lars Brink Professor Kwek Leong Chuan Professor Michael J Hannon Professor Phua Kok Khoo Professor Eliezer Rabinovici <u>Birmingham Organising Group</u> Professor Damien Walmsley (University of Birmingham) <u>Guests</u> Professor Nargiza Amirova (IAR Nagoya University)	**Mentor talks and responses** **Chair - Nargiza Amirova** **13:30-15:30** *Penny Andrews* *The "Casserole Constitution": The South African Constitution and International Laws* Response and discussion *Gil Kalai + two fellows* **15:30 Refreshments** **Chair - Lars Brink** **15:45-17:45** *Gil Kalai* *The Laws of Physics vs. the Laws of Computation – Laws and Errors in the Case of Quantum Computers* Response and discussion *Atul Parikh + two fellows*	**Evening** **Dinner** **Catering @ Orchid Room** **18:30** **20:00-21:00 Evening discussion** All fellows Penny Andrews Gil Kalai

4

Wednesday 21 March	Morning	Afternoon	Evening
Fellows All Fellows	**Mentor talks and responses** **Chair - Robin Mason** **09:00-11:00** *Partha Dasgupta* *A Matter of Trust: Laws and Norms* Response and discussion *David Gross + two fellows* **11:00-11:15** Refreshments **Chair - KK Phua** **11:15-13:15** *David Gross* *The Nature of Laws and Principles in Physics* Response and discussion *Patrick Geary + two fellows*	**Mentor talks and responses** **Chair - Mike Hannon** **14:15-16:15** *Michel Spiro* *CERN a Collective Machinery to Test Laws against Nature* Response and discussion *Michal Feldman + two fellows* **16:15-16:30** Refreshments **16:30-18:30** *Suzanne Blier* *From Fair Use to Zoning Ordinance: Challenging the Law for Public Benefit* Response and discussion *Partha Dasgupta + two fellows* **Dinner transport leaves at 19:00**	**Dinner** **20:00** Tanglin Club Churchill Room 5 Stevens Rd, Singapore 257814 http://www.tanglinclub.org.sg/abo ut/club-history.html
<u>Mentors</u> Professor Penny Andrews Professor Suzanne Blier Professor Sir Partha Dasgupta Professor Michal Feldman Professor Patrick Geary Professor David Gross Professor Gil Kalai Professor Atul Parikh Professor Michel Spiro <u>Directors</u> Professor Lars Brink Professor Kwek Leong Chuan Professor Michael J Hannon Professor Phua Kok Khoo Professor Eliezer Rabinovici <u>Birmingham Organising Group</u> Professor Damien Walmsley (University of Birmingham) <u>Guests</u> Professor Robin Mason (PVC International, University of Birmingham)			

5

Thursday 22 March	Morning and Afternoon	Evening
Fellows All Fellows	**Thematic panel day - to be planned by Fellows**	**Dinner hosted by Professor Robin Mason (PVC International University of Birmingham)**
Mentors Professor Penny Andrews Professor Suzanne Blier Professor Sir Partha Dasgupta Professor Michal Feldman Professor David Gross Professor Patrick Geary Professor Gil Kalai Professor Atul Parikh Professor Michel Spiro	09:00-10:30 SESSION 1 10:30-11:00 Refreshments 11:00-13:00 SESSION 2 13:00-14:00 Lunch **Lunchtime discussion** All fellows David Gross Michel Spiro	#06-01 National Gallery Singapore 1 St. Andrew's Road Singapore 178957 Tel: +65 9234 8122 Fax: +65 63845575 18:00 Welcome Drink Smoke and Mirrors 18:30-20:30 Dinner, Yan
Directors Professor Lars Brink Professor Kwek Leong Chuan Professor Michael J Hannon Professor Phua Kok Khoo Professor Eliezer Rabinovici	14:00-16:00 SESSION 3 **Includes 15:00 Refreshments break** Transport to leave at 17:00	Nanyang Executive Centre 23:00 – 23:30 (12:00 Sao Paulo Time) Link with UBIAS Directors' Meeting
Birmingham Organising Group Professor Damien Walmsley University of Birmingham		Session UBIAS programs: Results of ICA 1 and 2 Contact with ongoing ICA 3
Guests Professor Robin Mason (PVC International University of Birmingham)		

Friday 23 March	Morning	Afternoon	Evening
Fellows All Fellows Mentors Professor Penny Andrews Professor Suzanne Blier Professor Sir Partha Dasgupta Professor Patrick Geary Professor Wang Gungwu Professor Gil Kalai Professor Atul Parikh Professor Ernst Rank Professor Ada Yonath Directors Professor Lars Brink Professor Kwek Leong Chuan Professor Michael J Hannon Professor Phua Kok Khoo Professor Eliezer Rabinovici Birmingham Organising Group Professor Damien Walmsley (University of Birmingham)	**Mentor talks and responses** **Chair - Mike Hannon** **09:00-11:00** *Ernst Rank* *Laws and Beauty – a view from Computational Sciences* Response and discussion *Penny Andrews + two fellows* **11:00-11:15** Refreshments **Chair - KK Phua** **11:15-13:15** *Wang Gungwu* *Chinese Historiography, on History as a Mirror of Government* Response and discussion *Ada Yonath + two fellows*	**Project and outcomes challenge** **14:15-16:00** Fellows planning project outcomes **16:00** Refreshments **16:15-17:45** Presentation of projects	**Dinner** **18:30** **BBQ outside Orchid Room** **20:00-21:00 Evening discussion** All fellows Suzanne Blier Partha Dasgupta

Saturday 24 March	Morning	Afternoon	Evening
Fellows All Fellows **Mentors** Professor Suzanne Blier Professor Sir Partha Dasgupta Professor Patrick Geary Professor Atul Parikh Professor Ernst Rank Professor Ada Yonath **Directors** Professor Lars Brink Professor Kwek Leong Chuan Professor Michael J Hannon Professor Phua Kok Khoo Professor Eliezer Rabinovici	**Chair - Kwek Leong Chuan** **09:00-11:00** Atul Parikh *At the Limits of Laws of Thermodynamics: From Single Molecules to Living Matter* Response and discussion *Suzanne Blier + two fellows* **11:00-11:15** Refreshments **Chair - Eliezer Rabinovici** **11:15-13:15** Ada Yonath *Thoughts about Functional Flexibility and Origin of Life: The Creation of Multi Tasks Complex Systems* Response and discussion *Ernst Rank + two fellows*	**Transport leaves at 14:30** **15:30-17:30** **ArtScience Museum** 6 Bayfront Ave, Singapore 018974 Gather at 17:30 for dinner transport	**Dinner** **18:30** Buffet dinner Hotel Jen @ Orchardgateway 277 Orchard Rd, Singapore 238858 **Return transport to NEC at 20:30**

Sunday 25 March	Morning	Afternoon	Evening
Fellows All Fellows **Mentors** Suzanne Blier Patrick Geary Atul Parikh Ernst Rank Ada Yonath **Directors** Kwek L Chuan Michael J Hannon Phua Kok Khoo Eliezer Rabinovici Lars Brink	**08:00 Depart from NEC** Pulau Ubin and Chek Jawa Lunch at Pulau Ubin **14:30 Return to Singapore main island**	**15:00-15:45** Changi Museum **16:15-17:00** Sun Yat Sen Nanyang Memorial Hall	**Dinner** **17:30 Satay by the Bay** 18 Marina Gardens Drive #01-19 Singapore 018953 **Return transport to NEC at 20:30**

Monday 26 March

	Morning	Afternoon	Evening
Fellows All Fellows **Mentors** Professor Patrick Geary Professor Atul Parikh Professor Ernst Rank Professor Ada Yonath **Directors** Professor Lars Brink Professor Kwek Leong Chuan Professor Michael J Hannon Professor Phua Kok Khoo Professor Eliezer Rabinovici	**Fellows-led programme** 09:00-10:30 SESSION 1 10:30-11:00 Refreshments 11:00-13:00 SESSION 2 13:00-14:00 Lunch	**Fellows-led programme** 14:00-15:30 SESSION 3 15:30-15:45 Refreshments 15:45-17:30 SESSION 4	**Dinner** **Catering @ Orchid Room** **18:30** **20:00-21:00 Evening discussion** All fellows Atul Parikh Ernst Rank Ada Yonath

Tuesday 27 March

	Morning	Afternoon	Evening
Fellows All Fellows **Mentors** Professor Atul Parikh Professor Ernst Rank Professor Ada Yonath **Directors** Professor Kwek Leong Chuan Professor Michael J Hannon Professor Phua Kok Khoo Professor Eliezer Rabinovici	**Fellows-led programme** 09:00-10:30 SESSION 1 10:30-11:00 Refreshments 11:00-13:00 SESSION 2 13:00-14:00 Lunch	**Plenary and close** 14:00-15:00	

Eliezer Rabinovici

NANYANG TECHNOLOGICAL UNIVERSITY
Institute of Advanced Studies

INTERCONTINENTAL ACADEMIA

UNIVERSITY OF BIRMINGHAM
THE INSTITUTE OF ADVANCED STUDIES

Intercontinental Academia Birmingham 19 to 27 March 2019
Composers' Suite Edgbaston Park Hotel

Pre-Programme	14:00–16:00	17:30–18:30
Monday 18 March	*Perspectives in Chemistry:* ***From Supramolecular Chemistry towards Adaptive Chemistry*** **Professor Jean-Marie Lehn** Haworth 101 (Y2 campus map) Introduction Mike Hannon Director IAS	*Human Well-Being and Economic Accounting* **Sir Partha Dasgupta** Sir Alan Walters Harvard Lecture Theatre (R29 campus map) Introduction Professor Robin Mason PVC International

Monday 18 March — 18:30 Welcome drink and dinner Composers' Suite Edgbaston Park Hotel

Tuesday 19 March	Morning	Afternoon	Evening
Mentors and Directors Penny Andrews Lars Brink Partha Dasgupta Patrick Geary Mike Hannon Ernst Rank Eliezer Rabinovici Bo Rothstein Jean Marie Lehn David Gross **Guests** Prof. Hiroko Takeda Institute for Advanced Research, Nagoya University	09:00–09:15 Welcome Mike Hannon, Eliezer Rabinovici, Sue Gilligan 09:15–11:15 ***Steps Towards Life: Chemistry!*** Professor Jean-Marie Lehn Response Patrick Geary, Nkatha Kabira, Willem Gravett Chair: Mike Hannon **11:15 Group Photo 1** and refreshments 11:45–12:30 Director Talk Eliezer Rabinovici ***SESAME: A Source of Light for the Middle East*** 12:45–13:30 Director Talk Mike Hannon ***A Duet Between Chemistry and Biology*** 13:30–14:30 LUNCH outside Composers' Suite	14:30–15:45 Introduction to Fellows' Projects and Feedback from Mentors • ***Smart Cities, Dumb Laws*** • ***Ransomware: an interdisciplinary exercise*** • ***Interdisciplinarity*** • ***Human Laws, Machine Laws and the Meeting Points*** 15:45–16:00 Refreshments 16:00–17:15 ***Experiences of working on multidisciplinary projects.*** Penny Andrews, Ernst Rank Tom Schonberg and Irina Kuznetsova Chair: Eliezer Rabinovici 17:15–18:15 Pre-dinner UoB Brexit Briefing Professor Raquel Ortega-Argilés, Professor Anthony Arnull, Professor Scott Lucas, Dr Matt Cole, Dr Ben Warwick Chair: Mike Hannon	20:00 Dinner Edgbaston Park Hotel Restaurant After Dinner Mentor Slot 21:00–22:00

2

Wednesday 20 March	Morning	Afternoon	Evening
Mentors and Directors Penny Andrews Lars Brink Kwek Leong Chuan Partha Dasgupta Patrick Geary (out 12:15-15:00) Mike Hannon Atul Parikh (late afternoon) Eliezer Rabinovici (out 13:00-15:00) Ernst Rank Bo Rothstein David Gross (until 17:00) Ada Yonath (late afternoon) Guests Hiroko Takeda	08:45-10:30 Project Session 1 10:30-10:45 Refreshments 10:45-11:45 ***Understanding Constraints on Energy Policy*** Professor Martin Freer, Head of Physics and Astronomy Director of the Birmingham Energy Institute Chair Mike Hannon 11:45-13:00 ***Experiences of working on multidisciplinary projects*** Partha Dasgupta, Prof. Hiroko Takeda VV Binoy Misha Smolkin Anupam Chattopadhyay Chair Lars Brink 13:00-14:00 LUNCH Hotel Restauranrt	14:00-15:00 Project Session 2 15:00-17:00 ***Experiences of working on multidisciplinary projects*** Patrick Geary David Gross Isabel Kusche Uli Heisserer Chair Kwek Leong Chuan	19-30 Transport from hotel 20:00 Dinner Al Faizal's 136 - 140 Stoney Ln, Birmingham B12 8AQ
Outside the main programme	13:30-14:30 Seminar Patrick Geary, IAS Princeton *Bringing Genomic Data into Historical Research* Poynting Small Lecture Theatre S06 (R13 campus map)	13:30-14:30 Seminar Eliezer Rabinovici, (HUJI) *On the Phases of Gravity* Physics West 117 (R8 campus map) 18:00-19:00 **Peierls Lecture David Gross** *The Frontiers of Fundamental Physics* Poynting Lecture Theatre S02 (R13 campus map) Introduction Professor Andy Schofield PVC Head of College of Engineering and Physical Sciences	

Thursday 21 March	Morning	Afternoon	Evening
Mentors and Directors Penny Andrews (16:00 to BLS) Lars Brink (our 14:00-16:00) Kwek Leong Chuan Partha Dasgupta Patrick Geary Mike Hannon Atul Parikh (out 12:30-16:30) Ernst Rank (until 3pm)	09:00-11.00 *Recorded lecture* ***Making Sense of Corruption*** Professor Bo Rothstein Followed by Q. and A by Skype Response David Gross, Carla Ventura, Tom Schonberg Chair Eliezer Rabinovici **11:00 Group Photo 2** Refreshments	14:00-16.00 Project Session 3 Smart Cities Dumb Laws Panel 16:00-16.15 Refreshments 16:15-18.00 Pre-Dinner After Dinner Mentor Slot	20:00 Dinner Edgbaston Park Hotel Lloyd Room Hortnon Grange Dancing with DJ Hamann

Eliezer Rabinovici (out 12:30-14:00) Bo Rothstein (out 15:00-18:00) David Gross Ada Yonath (until 16:00) Guests Hiroko Takeda	11:30-12:30 *Experiences of working on multidisciplinary projects* Atul Parikh Petra Liedl Al Wilson Chair Mike Hannon 12:30-13:00 Fellows' Planning time 13.00-14.00 LUNCH Outside Composers' Suite	
Outside the main programme 13.00 Mentor and Director Lunch Host Professor Sir David Eastwood 12th Floor Muirhead Tower (R21 campus map) **Penny Andrews, Lars Brink** **Åsa Brink, Kwek Leong Chuan** **Partha Dasgupta,** **Patrick Geary, David Gross,** **Jacquelyn Savani, Atul Parikh,** **Eliezer Rabinovici, Ernst Rank** **Hiroko Takeda** **Geoffrey West, Ada Yonath**	15:00-16:00 **Lars Brink** *Alfred Nobel and the Nobel Prizes* Physics West 117 (R8 campus map) 15:00-16:00 **Atul Parikh** *Mixing Water, Transducing Energy,* *Shaping Membranes: Far from Equilibrium Biological Interfaces* Biosciences Room 301 (R27 on campus map)	17:30-18:30 **Ada Yonath** *Inaugural Grace Frankland Memorial Lecture* Haworth 101 (Y2 campus map) Introduction Professor Laura Green Pro-Vice-Chancellor and Head of College of Life and Environmental Sciences 18:00-19:00 **Penny Andrews** Inaugural Race and the Law Lecture *Race Inclusion and Excellence in Law Schools:* *A Perspective From Three Continents* Birmingham Law School Lecture Theatre 2 (R1 on campus map)

Friday 22 March	Morning	Afternoon	Evening
Mentors and Directors <u>Penny Andrews</u> Lars Brink Kwek Leong Chuan Partha Dasgupta (leaves evening) Patrick Geary Mike Hannon Atul Parikh Eliezer Rabinovici Bo Rothstein Michel Spiro (arrives afternoon) David Gross <u>Guest</u> Hiroko Takeda	08:30 Transport leaves EPH *09:30 Ada Yonath Interview about research. Hotel.* *In discussion with Bryan Turner* 10:00 Arrive Stratford Upon Avon Michel Saint Denis 10:15-11:15 <u>Professor Michael Dobson</u> As You Like it and Stratford Upon Avon 11.30-12.30 Cathleen McCarron (Voice Coach As You Like it) Group Voice and Text Session 11.30 Interview Professor Gross for University 12:30 Lunch	13:30-15.30 *Searching for the Laws of Life, Growth & Death from* *Organisms & Ecosystems to Cities & Companies* Geoffrey West Response Penny Andrews Katrien Pype Hanjo Hamann <u>Chair Kwek Leong Chuan</u> 15:30 Break Free time in Stratford Upon Avon	17:00 Early Dinner Rooftop Restaurant 19:15 As You Like It Royal Shakespeare Theatre Transport back to EPH

4

	Morning	Afternoon	Evening
Saturday 23 March Mentors and Directors Lars Brink Kwek Leong Chuan Patrick Geary Mike Hannon Atul Parikh Eliezer Rabinovici Bo Rothstein Michel Spiro David Gross Guests Hiroko Takeda	**Morning off** **If required transport will be arranged from the hotel** **12:30** Meet for lunch at the IKON Gallery 1 Oozells Square Brindleyplace, Birmingham B1 2HS	14:00-17:30 City walk to include Birmingham City Art Galleries Birmingham Library Gas Street Basin **If required transport will be arranged from the hotel to Pasta Di Piazza.**	20:00 Dinner Pasta Di Piazza 11 Brook Street St Paul's Square Birmingham B3 1SA
Sunday 24 March Mentors and Directors Lars Brink Kwek Leong Chuan Mike Hannon Eliezer Rabinovici Bo Rothstein (departs afternoon) Michel Spiro Guests Hiroko Takeda Dr Yukinori Kawae Graduate School of Law, Nagoya University	**Morning off** 13:00 Traditional Sunday Lunch The Plough	Fellow's Led Cultural or Work Programme	Light supper to be ordered individually at the hotel bar. Fellows' Project Session 4
Monday 25 March Mentors and Directors Lars Brink Kwek Leong Chuan Mike Hannon Eliezer Rabinovici Michel Spiro Geoffrey West Guests Hiroko Takeda Yukinori Kawae Professor Ary Plonski IEA USP Raouf Boucekkine Aix-Marseille University	09:00-10:45 Fellows' Project Session 5 10:45-11:00 Refreshments 11:00-13:00 Fellows' Project Session 6 13:00-14:00 LUNCH Outside Composers' Suite	14:00-15:00 *Experiences of working on multidisciplinary projects* Michel Spiro Geoffrey West Matsuo Yoh and Wai Yip Ho Chair Yukinori Kawae 15:00-15:15 Refreshments 15:15-15:45 Professor Bill Chaplin *Art & Science, in collaboration* 16:00-18:00 Chair Ary Plonski ICA Alum ICA 4	20:00 Dinner Edgbaston Park Hotel Restaurant After Dinner Mentor Slot 21:00-22:00

Tuesday 26 March	Morning	Afternoon	Evening
Mentors and Directors Lars Brink Kwek Leong Chuan Mike Hannon Eliezer Rabinovici Michel Spiro (until 12:30) Geoffrey West (until 15:30)JP **Guests** Yukinori Kawae Hiroko Takeda Ary Plonski Raouf Boucekkine	09:00-10:45 Fellows' Project Session 7 10:45-11:00 Refreshments 11:00-13:00 Fellows' Project Session 8 13.00-14.00 LUNCH Rear Bar area terrace	14:00-15:45 Fellows' Project Session 9 15:45-16:00 Refreshments 16:00-17:30 Fellows' Project Session 10	Dinner Edgbaston Park Hotel to be ordered individually at the hotel bar. After Dinner Mentor Slot 21:00-22:00
Outside the main programme	13:00-14:30 **Michel Spiro** *CERN: A collective machinery to test laws against nature* Aston Webb G33 (R4 on campus map)	16:00-17:30 **Geoffrey West** *A Physicist's Search for Simplicity and Unity in the Complexity of Living Systems from Cells and Cities to Ecosystems and Companies* Physics West 117 (R8 on campus map)	

Wednesday 27 March	Morning	Afternoon	Evening
Mentors and Directors Lars Brink Kwek Leong Chuan Mike Hannon Eliezer Rabinovici Michel Spiro Geoffrey West **Guests** Dr Yukinori Kawae Graduate School of Law, Nagoya University Prof. Hiroko Takeda Institute for Advanced Research, Nagoya University	09:00-10:45 Fellows' Project Session 11 10:45-11:00 Refreshments 11:00-13:00 Fellows' Project Session 12 13.00-14.00 LUNCH Outside Composers' Suite	14:00-15:00 Plenary Summary of ongoing projects & planning for next steps Mike Hannon Eliezer Rabinovici **CLOSE**	

Chapter 1

The "Casserole" Constitution: The South African Constitution and International Law

Penelope Andrews

New York Law School

1. Introduction

When I received the invitation in May 2017 to attend the 3rd UBIAS Intercontinental Academia at Nanyang Technological University, Singapore & University of Birmingham, I pondered on the meaning of the theme, namely *Laws: Rigidity and Dynamics,* and what input from South Africa I could bring to the discussions.

As the letter of invitation noted, *the concept of laws has different meanings in different settings, cultures and to individuals, which demands reflection, analysis and comparison, to elucidate the origins of laws and tensions among them.* The invitation letter also noted that when *examining laws devised by humans, natural laws or the evolution of theory emerging from chaos and complexity into laws, there is potential for intense interactions from Medicine to Linguistics, Chemistry to Music, History to Physics, Psychology to Economics, Engineering to Theatre and beyond.*

The question for me was how to fit into the theme the *intense interactions* that emerge from the complexity of apartheid and authoritarianism to the laws of democracy and social justice in South Africa. I chose to use the metaphor of a casserole, suggesting a mixture and a blending of different laws and legal traditions, consisting of the indigenous, national and international.

The design and emergence of the late 20th century constitutional project in South Africa provided an exciting global and national moment to recalibrate and consider the possibilities and limitations of a constitutional project purposely designed to generate particularized goals of democracy, equality and dignity, as pronounced in the Preamble to the Constitution:

- to heal the divisions of the past and establish a society based on democratic values, social justice and fundamental human rights;
- to lay the foundations for a democratic and open society in which government is based on the will of the people and every citizen is equally protected by law;
- to improve the quality of life of all citizens and free the potential of each person; and
- to build a united and democratic South Africa able to take its rightful place as a sovereign state in the family of nations.

The expansive South African Constitution directs courts to consider international and foreign law in their deliberations. Indeed, the impact of international and foreign law, especially in the detailed listing of a range of civil, political, social, economic and cultural rights, as well as the structures chosen to interpret and implement constitutional provisions, are quite apparent. The influence of international law, and to a lesser extent, foreign law, is also visible in the jurisprudence of the Constitutional Court.

The drafting of the constitution was an international effort. It was also of significant global import in the wake of several decades of global human rights advocacy and the ascent of a universal consensus around human rights, justice and the rule of law. Some, like the scholar Makau Mutua, have argued that,

...the construction of the post-apartheid state represented the first deliberate global effort to craft a human rights state, one animated by values of human rights, social justice and good governance.[a]

The transition from authoritarianism and apartheid to constitutional democracy in South Africa came at the end of a decades long global struggle to end apartheid. Arguably the global anti-apartheid movement was the most significant human rights movement of the late 20th century. Louis B. Sohn and Henry Richardson, both eminent international law

---

[a] Makau Mutua, "Hope and Despair for a New South Africa: The Limits of Rights Discourse", *Harvard Human Rights Law Journal* **10**, 63, 65 (1997).

scholars, have noted how the struggle against apartheid influenced and impacted the development of international law, especially then evolving international principles aimed at eliminating racism and apartheid.[b] Indeed, with the passage of the International Convention on the Suppression and Punishment of the Crime of Apartheid in 1973, apartheid was declared a crime against humanity.

In my essay I reflect on the impact and the role of international law in the drafting of the South African constitution as well as in the jurisprudence of the Constitutional Court. I also address the challenges of incorporating international law, especially international human rights law, into the constitutional project, and in particular the possibilities for generating a vigorous democracy and wide respect for human rights through such incorporation. The central question that I raise is whether universality as a value, norm or consensus, as reflected as part of the international law corpus, is sufficient to impel compliance, especially at the local level. Implicit in my investigation is an examination of the jurisprudential choices made by the Constitutional Court as it balances international law approaches as opposed to narrower localized interpretations of the Constitution.

This essay is divided into four sections: After this introductory section, the first section looks at the end of Apartheid and the transition to constitutional democracy. The second section examines the creation of the new constitutional dispensation with the drafting of the expansive Constitution and its detailed listing of rights, constitutional bodies and processes. The third section analyses the jurisprudence of the South African Constitutional Court and its engagement with international law in the interpretation of the Constitution. The final section outlines the challenges facing the pursuit and implementation of the rights embodied in the constitution, in light of the several legal and extra-legal institutional impediments, including the legacy of colonialism and apartheid, economic inequalities, persistent cultures of masculinities and an overall fragile culture of human rights.

---

[b] Louis B. Sohn, *Rights in Conflict: The United Nations and South Africa* (1994); Henry J. Richardson, "Self-Determination, International Law and the South African Bantustan Policy", *Columbia Journal of Transnational Law* **17**, 185 (1978).

2. End of Apartheid and the Transition to Constitutional Democracy

In 1948 the United Nations was established, reflecting a distinct global moment of optimism regarding human rights, peace and security. In the same year, pursuing a different path, the Nationalist Party, a party built on a political platform of white supremacy, came to power in South Africa. They hastily and enthusiastically embarked on a Kafkaesque project of racial segregation to separate the citizens of South Africa according to clearly demarcated racial groups.[c]

Apartheid was implemented at the same time that the United Nations Charter was adopted by the international community of states and negotiations were under way for the drafting of the Universal Declaration of Human Rights. Subsequently, apartheid

---

[c] In pursuit of this goal, the apartheid government passed a series of statutes to institutionalize racial discrimination. These statutes included: Population Act of 1950; Prohibition of Mixed Marriages Act of 1949; Group Areas Act of 1950; The Reservation of Separate Amenities Act of 1959. See *International Defence and Aid Fund, Apartheid: The Facts* (1982).

evolved as a contrast to the principles underlying the human rights declaration and, as mentioned earlier, was deemed as a crime against humanity by the United Nations in 1973.

The apartheid system violated the most basic tenets of international human rights law and policy, embodying a harsh combination of state-sponsored authoritarianism, militarism, race and gender discrimination, and economic exploitation. Therefore, the South African legal system was always of great concern to the international community, with some international legal scholars arguing that apartheid was a key factor in the development of international law as it relates to the principle of non-discrimination and state sovereignty.

It was the function of the apartheid legal system to bolster the grand plan of spatial and other forms of racial segregation. Laws and policies were designed to ensure that all South Africans would be allocated clearly defined roles within the racial context of white superiority and black inferiority. All aspects of people's lives, including where they worked, the positions they were entitled to hold in the workplace, whom they would marry, where they would live, which schools and universities they would attend, when and for what periods they could travel, were rigidly regulated. A highly sophisticated bureaucratic infrastructure, with a labyrinth of laws and regulations were designed to reinforce the inhumanity and brutality of apartheid, while maintaining the façade of effective and efficient governance for the White population, who constituted a demographic minority in South Africa.[d] A brutal security and police legal and extra-legal apparatus was constructed to ensure that apartheid laws were obeyed and that political dissent was stifled.

A cursory reading of the report of the South African Truth and Reconciliation Commission (TRC) vividly illustrates the grotesque lengths to which the minority White Apartheid government went to ensure that the system of apartheid, and effectively a global outcast among the family of nations, was reinforced. The TRC report documents in great detail not just the individual crimes against humanity and the gross violations of human rights, but also the institutional support for the system of apartheid, including the legal system, the medical

---

[d] *Ibid.*

establishment, the media, the religious institutions, and the business community.[e]

As I have noted above, the anti-apartheid struggle was one of the most important human rights struggles of the late 20th century, and Nelson Mandela arguably the most globally popular and revered of national leaders. When he died, Bishop Tutu, who himself was a revered figure of the anti-apartheid movement, wrote that:

> *Never before in history was one human being so universally acknowledged in his lifetime as the embodiment of magnanimity and reconciliation as Nelson Mandela was.*[f]

The release of Nelson Mandela from the infamous Robben Island became a rallying cry that animated a global movement. Indeed, by the late 1980s the combination of internal political opposition (especially from civil society, the religious community and trade unions), emboldened liberation movements in exile, and the isolation of South Africa through economic, sports, cultural and educational boycotts, led to its ultimate demise. On February 2, 1990 then President F.W. De Klerk announced the release of Nelson Mandela, the unbanning of the liberation movements including the African National Congress, the Pan Africanist Congress and the Communist Party, as well as other banned organizations. A state of emergency which had existed for several years was lifted and President De Klerk ordered the release of all political prisoners.[g]

These were tectonic shifts in a country that had been the poster child for racism and authoritarianism, the pariah of the world, and which had been torn about by violence in the years leading up to the transformative changes announced by President F.W. De Klerk.

The process of negotiations between the apartheid government and opposition parties and the liberation movements began in earnest in

---

[e] *Final Report of the Truth and Reconciliation Commission* (1998).

[f] Archbishop Desmond Tutu, "Nelson Mandela: A Colossus of Unimpeachable Moral Character", *The Washington Post*, December 6, 2013.

[g] Alister Sparks, *Tomorrow is Another Country: The Inside Story of South Africa's Negotiated Revolution* (1996).

1990. The negotiations led to the establishment of a Convention for a Democratic South Africa (from December 1991 to May 1992) and the Multi Party Negotiations Process, which culminated in the drafting of an interim constitution in 1993. South Africa's first elections were held in April 1994, the day that the Interim Constitution also came into force. Nelson Mandela was overwhelmingly elected as the first President of a democratic South Africa in an election watched by the global community.[h]

A Constitutional Assembly was mandated to draft the final Constitution guided by a set of constitutional principles incorporated in the 1993 Constitution. The 1996 Constitution was confirmed by the newly-established Constitutional Court in 1995.[i]

In 1993 Nelson Mandela and F.W. De Klerk, the last President of apartheid South Africa, jointly won the Nobel Peace Prize.

---

[h] *Ibid.*

[i] For an interesting discussion of the constitutional drafting process, see Hassen Ebrahim, *The Soul of a Nation: Constitution Making in South Africa* (1999).

3. Constitutional Structure and Constitutional Rights

The end of apartheid in South Africa came with the establishment of a
constitutional democracy in 1994. In its founding provisions, the new
Constitution outlines the human rights principles on which the new
democratic state is premised, including nonracialism, non-sexism, and
human dignity. The South African Constitution reflects the influence of
the global human rights framework and is a by-product of that
framework. For example, the Constitution embraces international law in
several ways, and the Constitution's comprehensive Bill of Rights is
drawn entirely from several human rights instruments, including the
Universal Declaration of Human Rights, the International Covenant on
Civil and Political Rights, the International Covenant on Economic,
Social, and Cultural Rights, the Convention on the Elimination of All
Forms of Discrimination Against Women and the Convention on the
Elimination of Racial Discrimination. The South African Bill of Rights is
expansive, incorporating a wide range of civil and political rights, as well
as economic, social, and cultural rights.

The second way that the Constitution incorporates international
law is that it specifically directs the South African courts to consider
international law in their deliberations. In addition, it provides for the
direct incorporation of international law into the South African legal
system. South Africa is party to several international human rights
instruments that include the elimination of racial discrimination,
slavery, and genocide; the suppression of human trafficking; and the
rights of women, children, and refugees.

In the Founding Provision of Chapter 1, the Constitution makes clear
that the South African state is founded on several values, including
"[h]uman dignity, the achievement of equality and the advancement of
human rights and freedoms, [n]on-racialism, and non-sexism." The most
important provisions relating to equality are found in the Bill of Rights,
particularly Section 9, which outlines the principle of equality in some
detail. Section 9 contains a general commitment to equality before the
law and equal protection of the law, dictating that:

The state may not unfairly discriminate directly or indirectly against anyone on one or more grounds, including race, gender, sex, pregnancy, marital status, ethnic or social origin, colour, sexual orientation, age, disability, religion, conscience, belief, culture, language and birth.[j]

Section 9 also provides that "no person may unfairly discriminate directly or indirectly against anyone on one or more grounds" as listed above.

Section 9 takes a further cue from international trends that expand the definition of equality by incorporating affirmative action. It specifically provides that:

Equality includes the full and equal enjoyment of all rights and freedoms. To promote the achievement of equality, legislative and other measures designed to protect or advance persons or categories of persons disadvantaged by unfair discrimination, may be taken.[k]

The Constitution also provides that "everyone has inherent dignity and the right to have their dignity respected and protected", as well as rights relating to "freedom and security of the person". The latter includes the right to be free from all forms of violence in either the public or private sphere. The Constitution also lists a range of rights and processes to ensure that everyone can access the rights embodied in the Constitution, as well as constitutional bodies to implement and enforce the rights. In addition to the Constitutional Court and other courts, several bodies are mandated to pursue the human rights embodied in the Constitution. These include the Public Protector, the Human Rights Commission, the Commission for Gender Equality, the Electoral Commission, and the Commission for the Promotion and Protection of the Rights of Cultural, Religious, and Linguistic Communities.

The Constitution also incorporates a range of economic, social and cultural rights that are placed on the same constitutional footing as civil and political rights, an approach that was designed to comprehensively address the poverty and economic inequalities that

---

[j] Chapter 9, *Constitution of South Africa* (1996).
[k] *Ibid.*

pervade South Africa. In this way the South African Constitution has attempted to bridge the traditional divide in international law between civil and political rights on the one hand and economic, social and cultural rights on the other, making both sets of rights equally legally significant and fully justiciable before the courts.

This comprehensive approach to constitutionalism has earned the Constitution great international acclaim. It is seen as a transformative document, one that may provide the legal space for overhauling the political, economic and social legacy of apartheid.

It is worth noting the international context and the global moment within which the drafting of the South African Constitution was initiated. The constitutional negotiations and the transition to democracy in South Africa occurred in the wake of the fall of the Berlin Wall in 1989 and the collapse of the communist regimes in Eastern Europe. This period was seen as the high water mark of capitalism, a triumph as it were, and the ascendancy of liberal democracy and the rule of law, enabling the so-called Washington Consensus. Indeed, as observed by Upendra Baxi, the Indian legal scholar, the language of human rights came to replace the language of economic distribution. Human rights discourse was seeking to "supplant all other ethical discourses".[l] In a similar vein, Boa de Sousa Santos, the Portuguese scholar, has noted how human rights became the lingua franca of "progressive politics," providing an "emancipatory script" for those seeking redress from injustice.[m]

This had somewhat of a political effect in South Africa, reverberating in the constitutional choices made, and as exemplified by the constitutional edifice. Although the national liberation movements had largely been committed to policies of economic redistribution, by the 1990s the international consensus had shifted to one in which human rights were characterized not by redistribution of material resources in the world, but in legal texts such as bills of rights.[n] The South African

---

[l] Upendra Baxi, "Voices of Suffering and the Future of Human Rights", *Transnational Law and Contemporary Problems*, 126 (1998).
[m] Boaventura de Sousa Santos, *Toward a New Legal Common Sense. Law, Globalization, and Emancipation* (2002).
[n] Upendra Baxi, *Voices of Suffering and the Future of Human Rights supra* Note 1.

Constitution and its expansive Bill of Rights reflects this paradigmatic shift in the characterization and articulation of human rights norms.

That the South African Constitution is a compromise document located within the context of the triumphalist ascendancy of a global liberal legal ideology at the end of the 20th century does not detract from its transformative potential. To discount this possibility and what it has meant for South Africans, especially black South Africans, is almost to ignore the potency of rights and dignity. The fundamental legacy of apartheid was a complete erasure — of humanity, of dignity, of all rights. The Constitution with its extensive listing of rights and its anchor in dignity is the foundation and the architecture to restore and give back to South Africans humanity, dignity and rights that were erased under apartheid.

As Patricia Williams noted so poignantly regarding a similar critique from critical legal scholars about civil rights law in the United States:

> *"Rights" feels new in the mouths of most black people. It is still deliciously empowering to say. It is the magic wand of visibility and invisibility, of inclusion and exclusion, of power and no power. The concept of rights, both positive and negative, is the marker of our citizenship, our relation to others.*°

In addition, the South African Constitution represents a vindication of decades of human rights activism, not just because of its expressed human rights commitment in the Bill of Rights, but also because the Constitution made South Africa, formally at least, a version of the penultimate human rights state.

But despite these limitations, if one looks at the trajectory of human rights discourse during the decades following the passage of the Universal Declaration of Human Rights, including those shameful periods when human rights became hostage to cold war politics, the South Africa embrace of human rights principles provided a welcoming ray of hope at the end of the 20th century. In the 30 years since South Africa became a democracy, and particularly in the wake of the so-called war on terrorism, the discourse of human rights has once more become

---

° Patricia J. Williams, *The Alchemy of Race and Rights* 164 (1991).

contested and distorted, often held hostage to opportunistic politics and an increasing global cynicism.[p]

BILL OF RIGHTS

EQUALITY
Everyone is equal before the law and may not be unfairly discriminated against.

HUMAN DIGNITY
Everyone has inherent human dignity which must be respected.

LIFE
Everyone has the right to life.

FREEDOM AND SECURITY OF THE PERSON
You have a right not be physically detained without trial or abused in any way.

SLAVERY, SERVITUDE AND FORCED LABOUR
You may not be subjected to slavery or forced labour.

PRIVACY
Your right to privacy includes your body, home and possessions.

FREEDOM OF RELIGION, BELIEF AND OPINION
You have the right to think, believe and worship.

FREEDOM OF EXPRESSION
You have the right to say, read and study whatever you choose but hate speech is not allowed.

ASSEMBLY, DEMONSTRATION, PICKET AND PETITION
You have the right to peacefully assemble, demonstrate and protest.

FREEDOM OF ASSOCIATION
You have the right to associate with anyone.

POLITICAL RIGHTS
You may form a political party, run for office and vote for any party in free and fair elections.

CITIZENSHIP
No citizen may be deprived of citizenship.

FREEDOM OF MOVEMENT AND RESIDENCE
You have the right to enter and leave the Republic at will.

FREEDOM OF TRADE, OCCUPATION AND PROFESSION
You have the right to choose any legal trade or occupation freely.

LABOUR RELATIONS
Every worker and employer has the right to organise and negotiate to further their aims.

ENVIRONMENT
You have the right to live in a protected, healthy environment.

PROPERTY
No-one may be deprived of property, except in terms of law of general application.

HOUSING
You have the right to have access to adequate housing.

HEALTH CARE, FOOD, WATER AND SOCIAL SECURITY
You have the right to have access to health care, adequate food and water and social security.

CHILDREN
Every child has the right to a name, nationality and protection from abuse and exploitation.

EDUCATION
You have the right to receive basic education in the official language of your choice where that education is reasonable practicable.

LANGUAGE AND CULTURE
You have the right to use the language of your choice and practise your own culture.

CULTURAL, RELIGIOUS AND LINGUISTIC COMMUNITIES
You have the right to form, join and maintain cultural, linguistic and religious grouping of your own choice.

ACCESS TO INFORMATION
You may access any information held by the state for the protection of your rights.

JUST ADMINISTRATIVE ACTION
You have the right to administrative action that is lawful, reasonable and procedurally fair.

ACCESS TO COURTS
You have the right to resolve your legal disputes in a court or another impartial tribunal.

ARRESTED, DETAINED AND ACCUSED PERSONS
When arrested for allegedly committed an offence, you have the right to remain silent, to be brought before a court within 48 hours and the right to legal representation

LIMITATION OF RIGHTS
Everyone's rights may be limited. The limitation should apply to everyone to the extent that it is reasonable and justifiable in an open and democratic society based on human dignity, equality and freedom.

RESPONSIBILITIES
All citizens are equally subject to the duties and responsibilities of citizenship.

ALL THESE LAWS ARE SUBJECT TO THE LAW OF THE LAND, BUT APPLY TO ALL WHO LIVE IN THE REPUBLIC OF SOUTH AFRICA.

the **doj &cd**
Department
Justice and Constitutional Development
REPUBLIC OF SOUTH AFRICA

STAND TOGETHER FOR RIGHTS

---

[p] Helen Duffy, *The 'War on Terror' and the Framework of International Law* (2015); Daniel Moeckli, Human Rights and Non-discrimination in the 'War on Terror' (2008).

4. Constitutional Jurisprudence

Underpinning constitutional interpretation in South Africa is the recognition that the adjudication of rights in the Constitution occurs in a societal context of deep economic and social inequalities. The task of judges is therefore to be mindful of their role as giving effect to the transformative goals of the Constitution.

The legacy of apartheid and the particular constitutional mandate (as the court of last resort on constitutional issues) meant that in the early years after its establishment, the Constitutional Court overwhelmingly adjudicated transitional issues of social justice and human rights. The expansive and comprehensive embrace of a range of civil, political, social, economic and cultural rights, as well as the incorporation of international human rights law, has resulted in constitutional interpretation in South Africa spawning a wide-ranging and pioneering jurisprudence on equality and dignity that continues to be cited in many jurisdictions.

The use of the death penalty, particularly against political opponents of apartheid and disproportionately against black males, was one of the first legacies of apartheid to be confronted by the Court. Invoking the right to life and the right to dignity found in the Bill of Rights and international human rights law, the court struck down the death penalty as unconstitutional.[q]

The Constitutional Court has also examined in some detail the issue of equality — the paramount principle in the Bill of Rights. The Court has formulated a substantive vision of equality by exploring a range of factual situations, including those involving the rights of HIV-positive persons not to be discriminated against in their employment; the right of prisoners to vote; the rights of unmarried fathers in relation to adoption of their children; the rights of permanent residents not to be treated unfairly in the workplace compared to citizens; the rights of homosexuals to engage in

---

[q] *S v Makwanyane and Another* 1995 (3) SA 391 (CC).

consensual sexual conduct; and the rights of African girls and women not to be discriminated against under indigenous customary law.

Addressing the issue of violence, one of the most fraught problems in South Africa, the Constitutional Court has held the state accountable for failure to protect citizens from harm caused by third parties, especially in public places. So, for example, the Court has held the national transport authority liable for not protecting train passengers from criminal conduct.[r] Regarding violence against women in both the public and the private spheres, the court has rendered judgments that reflect its constitutional commitment to eradicating violence against women, again holding the state liable for failure to protect women from third party perpetrators.[s]

The Constitutional Court has also attempted to strike a balance between competing sets of rights. So in evaluating the competing rights of privacy and state regulation, the South African Constitutional Court has deliberated on the often competing claims of religious rights on the one hand, and equality, on the other. It has also attempted to balance the rights of criminals in a violent society such as South Africa and the rights of individuals to personal security. So too the Court has balanced the scope and limits of criminal sanction against the rights of prisoners to exercise their right to vote while incarcerated.[t]

Of significance has been the court's incremental adoption of a socioeconomic rights jurisprudence that strives to grapple with the dire economic conditions of South Africa. Mindful of the doctrine of separation of powers and reticent to usurp the prerogative of the legislature, in addition to its concerns about institutional capacity, the Court has nonetheless attempted to ensure that the government pays attention to the needs of the poor. For example, in a landmark judgment in 2000, the court outlined in great detail the government's obligations to provide housing for those most in need of shelter.[u] It has also mandated that the government, in compliance with the right to

---

[r] *Mashongwa v PRASA* 2016 (3) SA 528 (CC).

[s] *Carmichele v Minister of Safety and Security* 2001 (4) SA 938 (CC);

[t] *August and Another v Electoral Commission and Others* 1999 (3) SA 1.

[u] *Government of the Republic of South Africa and Others v Grootboom and Others* 2001 (1) SA 46.

health as delineated in the Bill of Rights, provide antiretroviral drugs to HIV-positive pregnant women at public hospitals throughout South Africa.[v]

Regarding the rights of indigenous women and girls not to be discriminated against by indigenous law and custom, the Court has signaled clearly that indigenous laws and policies are only valid as long as they do not violate the right to equality and dignity for individual members of the community. Mindful of the odious history of colonialism and apartheid *vis a vis* indigenous communities, the Court has contextualized the rights to autonomy of indigenous communities and has articulated the need for respect of indigenous rights and the need to eradicate odious practices and attitudes towards indigenous peoples. This history is a particularly shameful one.[w] But the Court has also struck down as unconstitutional indigenous laws that discriminate against women and girls, as they did in a case involving the rights to inheritance denied to women and girls as a group.[x]

The effectiveness of the Constitutional Court in this human rights venture has been somewhat compromised by some of the lower courts' inabilities to fully effectuate the constitutional rights. Many of the lower courts, particularly those in the rural areas, are under resourced and lack properly trained and experienced personnel. In addition, although the Constitutional Court has rendered some pioneering decision regarding the implementation of social and economic rights, providing considerable hope to claimants, its judgements have not always been followed through by government officials. The Court is ultimately dependent upon the willingness of government officials and members of civil society to enforce its judgments.

For the most part, the court's judgments have been respected by both the government and society at large, although many South African citizens, including government officials, are still skeptical of the new human rights environment, particularly as it pertains to criminals, whom

---

[v] *Minister of Health v Treatment Action Campaign (TAC)* (2002) 5 SA 721 (CC).
[w] Martin Chanock, *The Making of South African Legal Culture 1902–1936: Fear, Favor and Prejudice* (2001).
[x] *Bhe v. Magistrate Khayelitsha & Others.* 2005 (1) BCLR 1 (CC).

they see as "having too many rights".[y] In addition, widespread xenophobia in South Africa, particularly against African immigrants and refugees, has put paid to the idea of South Africans embracing human rights with enthusiasm. The very progressive and generous Constitution and Bill of Rights arguably sits atop illiberal society.

As mentioned earlier, the South African government has sometimes been lacking in implementing the Constitutional Court's socioeconomic judgments. Such omissions result from a mixture of incompetence, hostility, or indifference. The success of the transformative constitutional and legal enterprise in South Africa is dependent upon a productive collaboration between the legislature, bureaucracy, and the courts, especially the Constitutional Court. This is especially crucial in the area of socioeconomic rights, where failure to act on or implement courts' judgments may discredit the entire constitutional endeavor.

In the final analysis, the limitations of vigorous enforcement of court judgments stem from a range of factors, including the lack of implementation and enforcement of rights due to absence of will, commitment, inertia, and incompetence on the part of government, the private sector and the wider society; lack of access to legal resources on the part of those whose rights ought to be pursued, especially their socio-economic rights; deeply ingrained illiberal philosophical ideas and attitudes, including racism, sexism, patriarchy, homophobia and xenophobia leading to the lack of a shared human rights culture across the entire society; and the widespread poverty and economic inequality that severely test the sustainability of the constitutional framework.

5. Challenges to Constitutional Democracy — South Africa and Globally

The possibility of widespread societal change through the mechanism of a constitution and Bill of Rights is severely constrained in the face of widespread poverty and massive economic and social inequalities. In

---

[y] Chantall Presence, *#CrimeStats: Criminals have too many rights, says Mbalula,* October 24, 2017 at: *https://www.iol.co.za/news/crimestats-criminals-have-too-many-rights-says-mbalula-11689907.*

South Africa, as is the case globally, the current global economic model has resulted in many being lifted out of poverty, but has left far too many people in dire circumstances. The pressure on the South African government to direct economic policy with the purposive goal of addressing poverty and economic inequality is urgent. Failing to do so threatens the constitutional framework and may lead to widespread disillusionment with law and legal processes.

A further challenge to the constitutional framework is the endemic corruption that has become part of the body politic in South Africa. The almost decade long leadership of President Jacob Zuma resulted in widespread looting of the state coffers, an institutionalization of state capture of key national resources, including state-owned enterprises, and the hollowing out of state institutions to serve the corrupt ends of the President and his allies. President Ramaphosa was elected 5 years ago, with the expectation that he would attempt to redress and reconfigure state economic policy. But these attempts on his part, albeit rather limited ones, have remained a challenge of Sisyphian dimensions. The President's failure to address these economic questions as a matter of urgency threatens the constitutional arrangements and may lead to cynicism about the possibilities of legal change.

Ubiquitous crime and lack of personal safety and security in South Africa, a consequence of a range of distinct and interlinking factors, has also resulted in a level of exhaustion and exasperation with the constitutional project, where the rights of the criminally accused are seen as impinging on the rights of victims of crime. This has created a response that mirrors some of the "law and order" calls that we witness in countries of the global north, particularly around the reception of migrants. Similarly, the alarming incidence of xenophobia has its roots in erroneous assumptions about the criminal proclivities of migrants. These attitudes severely test the constitutional commitment to human rights.

Finally, South Africa rates among the most misogynistic cultures with shocking statistics of violence against women. I have written elsewhere about the deep and sustained cultures of masculinities that pervade South African life, across class, race, culture, age and other identity markers.[z]

---

[z] Penelope E. Andrews, "Violence Against Women in South Africa: The Role of Culture and the Limitations of the Law", *Temple Civil and Political Rights Law Review* **8** (1998)

These cultures of masculinities present a formidable challenge to South Africa's constitutional project.

A combination of these factors, operating either individually or in concert with each other, pose substantial challenges to the constitutional project in South Africa. Many of these factors are not unique to South Africa, and are in fact mirrored elsewhere. For example, the widespread xenophobia that we are witnessing in the affluent countries of the global north, as well as economic inequalities and patriarchal cultures, means that the local South African problems require national as well as global attention.

South Africa: Two Countries

6. Concluding Comments

In conclusion, as I return to my opening remarks regarding the subject of the 3rd UBIAS Intercontinental Academia, the South African Constitution and Bill of Rights has drawn from the *concept of laws* which has *different meanings in different settings, cultures and to individuals,* but which has now become part of the constitutional fabric of the society. This embrace of an amalgam of values embodied in the

425; Penelope E. Andrews, "Learning to Love After Learning to Harm: Post-Conflict Reconstruction, Gender Equality and Cultural Values", *Michigan State University College of Law Journal of International Law* **15** (2007) 41.

law, the casserole as it were, has created the possibilities for a particular dynamism that will influence the implementation and enforcement of constitutional rights in South Africa. Conversely, it may also allow South Africa's **constitutional** project to influence and provide meaning elsewhere. If nothing else, it demonstrates the potential for *intense interactions* at the local and global level.

Chapter 2

Laws of History and Laws in History

Patrick Geary

Institute for Advanced Study

1. Introduction

As a practicing historian, rather than a philosopher who reflects theoretically on history, I am rather uneasy with the theme and parameters of this volume. When I first learned that the theme was to be "Laws: Rigidity and Dynamics," I assumed that the contribution that an historian might make to this discussion would be to look at how societies construct and use laws, norms, customs, and what more generally can be termed, after the French sociologist Pierre Bourdieu, habitus, that bundle of physical and mental modes of thinking, feeling, deciding, and being, that the individual acquires from his or her position in a social structure and employs to structure his or her decisions, actions, and life. Having accepted to participate, however, I subsequently learned that our organizer, Professor Eliezer Rabinovici, was thinking as a physicist when he proposed laws as our theme. He asked me to reflect on whether there are laws governing certain historical processes that one may dare to abstract; or is each process something very special?

Of course, even in physics there is a lively discussion of the role of laws and in particular how meaningful the search for a "final theory" actually is. But even as strong a proponent of a final theory as Stephen Weinberg admits that if one of the thousands of competing string theories should develop final physical principles that have no explanation in terms of deeper principles, these laws governing initial conditions "will never be able to eliminate the accidental and historical elements of sciences like biology and astronomy and geology."[1] The problems, as he points out, are first the possibility of chaos, even in simple systems. "The presence of chaos in a system means that for any given accuracy with which we specify the initial conditions, there will eventually come a time at which we lose all ability to predict how the system will behave."[2] The second problem, he writes, is that the intrusion of historical accidents set permanent limits on what we can explain. His third problem is "emergence": "As we look at nature at levels of greater and greater complexity, we see phenomena emerging that have no counterpart at the simpler levels."[3] This is certainly true in biology and, according to Weinberg, even in physics. Now it may be possible that a final theory will deal with chaos, historical accidents, and emergence, but while waiting for this we historians must muddle along as best we can.

The question of whether laws governed the course of history was for centuries easy to answer: For Jews and Christians, history was the unfolding of a divine plan: the laws of history were divine laws. Moreover, these laws could be comprehended by humans. However, by the eighteenth century, God was eliminated from history, and intellectuals sought alternative ways to understand and explain historical change. Were historical movements, events, and developments meaningful or were they random? Was there a teleology to history? These questions spoke to the ultimate value of studying the past: Did a study of the past inform people of the present about how to act at a personal, national, and societal level? Were there "lessons of history," in the sense that the past, properly understood, could be predictive of the future? If not, if history were merely a series of random unique events that defied explanation or meaning, then what possible value could be given to the study of the past? Was the past, in the words of Shakespeare, merely "a tale told by an idiot, full of sound and fury, signifying nothing?"

The heyday of investigation into whether laws govern historical processes was the nineteenth and first half of the twentieth century. Grand theories of history, beginning with that of the German philosopher Georg Wilhelm Friedrich Hegel (1770–1831) through Karl Marx (1818–1883), and Arnold J. Toynbee, (1889–1975), sought sweeping laws that explained macro-events in human history.

Hegel understood world history as "the necessary development, out of the concepts of mind's freedom alone, of the moments of reason and so of the self-consciousness and freedom of mind."[4] While individual people, states, and other entities may act from self-interest, actually all are instruments of the mind or Spirit of the world (*Weltgeist*). History, properly understood, could then be seen to be developing according to the specific dialectical principles of this Spirit. He distinguished realms of history, identified with specific civilizations chronologically organized. These were the Oriental, Greek, Roman, and Germanic, the latter being identified with Western Europe, and culminating in the modern European nation-state as the end or goal of history.[5] Such a schema made it possible to understand what might be termed laws governing the different realms as well as the process of self-realization in the European nation state.

Marxist historical materialism emphasized the primacy of material life over all forms of culture and the dependence of the latter on the former. While Karl Marx himself had a complex and subtle understanding of the historical process, in subsequent Marxist theoreticians, dialectical materialism and historical materialism were the same and increasingly reduced to a mechanistic process.[6] The material conditions of production fundamentally determine the organization and development of societies. These modes of production necessarily condition social, cultural, political, and intellectual life which are superstructure, entirely dependent on the infrastructure which are the material conditions of production. Ultimately, the material modes of production come into conflict with relationships of production — that is, the property relationships within which they have operated. The result is conflict and, ultimately, revolution which leads to a transformation of the superstructure. This system provides a way of understanding historical causes, laws if one will, since history is not the random succession of events or structures, but the systematic and predictable dialectic of class struggle firmly grounded in the means of production. Concentrating on Western Europe, Marx and Engels postulated six distinct stages of historical development, all directly dependent on (we might say caused by, or obeying the laws of) the material means of production of the age. These were primitive communism, slave society, feudalism, capitalism, socialism, and ultimately communism. Nevertheless, both Marx and Engels disavowed the idea that a great theory of the laws of history could explain the past or predict the future in detail. Rather, these laws were more the limitations and conditions within which the dialectical process might take place.

While Hegel had posited four chronologically and culturally distinct civilizations and Marx concentrated on the material conditions that led to six successive stages of history, Arnold Toynbee took a radically comparative approach to human history. In his massive twelve-volume *A Study of History*, he examined the rise and fall of 26 civilizations across human history, looking for what he explicitly termed laws that accounted for their rise and fall.[7] In reality, his "laws," are rather disappointing: As he stated it in 1934:

> We have ascertained that civilizations come to birth in environments that are unusually difficult and not unusually easy, and this has led us on to inquire whether or not this is an instance of some social law which may be expressed in the formula: 'the greater the challenge, the greater the stimulus'.[8]

He categorized five stimuli for civilizations: hard countries, new ground, blows, pressures and penalizations. Civilizations develop when these challenges are difficult but not overwhelming; they continue to grow by surmounting new challenges, led by creative minorities who inspire others to follow them. The decline of civilizations results from the transformation of this creative minority into a dominant minority that attempts to impose its control, not through inspiration but through coercion. Remarkably, this vast undertaking, riddled with inaccuracies, superficialities, and gaping lacunae, was enormously popular in the middle decades of the twentieth century with millions of copies of his works sold. After his death, however, Toynbee's work rapidly disappeared from view and is scarcely even given notice by contemporary historians.

I have little sympathy for any of these searches for universal laws governing history: They are too formulaic, too extraneous, and appear as vastly simplified *ex post facto* justifications of the present as the inevitable product of these laws. These kinds of "Laws of History" simply do not intersect with anything that I recognize as historical research or analysis.

A very different and more challenging defense of the laws of history has come, not from philosophers of history or meta-historical researchers like Toynbee, but from analytic philosophers. Their argumentation is ultimately derived from David Hume's analysis of causes. Hume (1711–1776) argued that when one posits a relationship of cause and effect, what one is actually asserting is simply that certain regularities of succession have been observed in the past. One cannot prove that one billiard ball striking another causes the second ball to move in a certain direction and with a certain velocity. All that one has actually observed is that repeatedly, one ball strikes the other in a certain way, and the other moves in a certain way. Repeated observations in the past lead to the unprovable assertion that the first ball is the cause of the second ball's motion:

> The idea of cause and effect is deriv'd from experience, which presenting us with certain objects constantly conjoin'd with each other, produces such a habit of surveying them in that relation, that we cannot without a sensible violence survey them in any other.[9]

The logical positivist Carl Hempel (1905–1997), developed Hume's analysis of causation further in his analysis of history. In a seminal 1947 article, he argued that if historians attribute causes to historical developments, then they are implicitly positing laws similar to those of the natural sciences.[10] His argument ran as follows: By law he meant "a statement of universal conditional form which is capable of being confirmed or disconfirmed by suitable empirical findings."[11] This sort of "law'" follows the general type: "In every case where an event of a specified kind C occurs at a certain place and time, an event of a specified kind E will occur at a place and time which is related in a specified manner to the place and time of the occurrence of the first event."[12] Of course, C need not be only one specific event.

Now the assertion that a set of events — say, of the kinds C_1, C_2, ... C_n — have caused the event to be explained, amounts to the statement that, according to certain general laws, a set of events of the kinds mentioned is regularly accompanied by an event of kind E.

He continues that a scientific explanation of an event would consist of the following:

(1) A set of statements asserting the occurrence of certain events C_1, C_n, at certain times and places,

(2) A set of universal hypotheses, such that

    (a)   The statements of both groups are reasonably well confirmed by empirical evidence,

    (b)   From the two groups of statements the sentence asserting the occurrence of event E can be logically deduced.

Number (1) presents the determining conditions for the occurrence of the event that one seeks to explain. Number (2) contains the general laws on which the explanation is based. They posit that whenever the boundary conditions (1) are present, event E will occur.

When applied to historical events, Hempel acknowledged that condition (1) can never be complete in the sense that "a statement of all the properties exhibited by the spatial region or the individual object involved, for the period of time occupied by the event in question,"[13] but nevertheless, while "it is impossible to explain an individual event in the sense of accounting for all its characteristics by means of universal hypotheses, … the explanation of what happened at a specified place and time may gradually be made more and more specific and comprehensive."[14] He is also aware that while a general regularity is posited in historical explanations, it may be impossible to judge the exact extent to which conditions existed that would inevitably result in specific effects.

Hempel argues that this is true in the natural sciences as well as in history and that there is no essential difference between the ways that the two develop explanations. He sees no essential difference between statements that are scientific predictions about future events and statements that describe present or past conditions, since both evoke universal empirical hypotheses. Historians cannot claim to explain historical phenomena, whether it is the cause of the assassination of Julius Caesar or the causes of the industrial revolution in Britain, unless general laws that connect cause and effect are implicitly or explicitly being applied. If historical explanations are intended to show that events did not occur simply by chance but were to be expected because of certain antecedent or simultaneous conditions, then the historian is implicitly claiming that there are general laws that determine historical developments. Of course, historians do not tend to state such general laws explicitly. However, Hempel goes on to say:

> Particularly, such terms as "hence," "therefore," "consequently," "because," "naturally," "obviously," etc., are often indicative of the tacit presupposition of some general law: they are used to tie up the initial conditions with the event to be explained; but that the latter was "naturally" to be expected as "a consequence" of the stated conditions follows only if suitable general laws are presupposed.[15] Thus the very language of historical narrative, in its use of adverbs, actually implies historical laws, however vehemently historians may wish to deny this.

At the same time, Hempel recognized that many historical explanations do not rely on laws in the form of universal conditionals, but are of a statistical nature, indicating only a probability. Because complete understanding of all preexisting conditions is not normally possible in historical research and analysis, Hempel characterizes the kind of explanatory analyses of historical events as an *explanation sketch*, that is, "Such a sketch consists of a more or less vague indication of the laws and initial conditions considered as relevant, and it needs "filling out" in order to turn into a full-fledged explanation."[16] This filling out involves a gradually increasing precision of the empirically verifiable formulations and, along with it, the rejection of non-empirical explanations such as appeals to "national character," "destiny," "empathy," or some metaphysical theory of history such as the Weltgeist or dialectical materialism.

To what extent do the musings of an analytic philosopher like Hempel correspond to the actual way that historians understand historical causation? Are in fact historians consciously or unconsciously positing laws when they talk of historical causation? In the remainder of my time, I would like to discuss how working historians actually understand and work with causation, questioning what if anything Laws might mean in what we actually do. Being an empirical historian, I will do this by exploring some current cases in which cause, and thus potentially laws loom large in historical debate.

The first and in some ways the most straight-forward case is in the increasingly active field of environmental history. Historians have long struggled to explain the extraordinary rise, in the course of the thirteenth century, of the Mongol Empire. In the twelfth century, the Mongolian plateau was dominated by various nomadic tribal confederations, among them the Mongols, who were defeated and held in check by their Tatar neighbors and the Chinese Jin emperors. Everything changed with the rise of Temujin or Chinggis Khan (1162–1227) who defeated rival Kahns, united the Mongols, and then rapidly conquered Central Asia and much of China. Under his immediate successors, the Mongol Empire ultimately extended from the Pacific well into Europe, the largest empire that the world has ever known. How was it possible for a loosely organized confederation of nomadic shepherds and herdsmen to create the greatest empire in history in under a century? Since the only natural resource in the Mongolia Steppe is its grasslands, its pastoral economy is directly linked to growing season moisture availability. Thus the environment has long been seen as a crucial factor in understanding the rise of the Mongol Empire. But how? One favored hypothesis was that severe drought had pushed the Mongols out of their accustomed territories in search of grazing ground, a process that set in motion this extraordinary conquest. Another, alternative hypothesis was that a long period of favorable climate produced an abundance of grass, allowing Steppe populations to greatly increase the size of their herds of horses and thus develop a more powerful military. Only recently has scientific evidence been produced that might support one or the other hypothesis. Recently a team composed of my colleague Nicola Di Cosmo, and a group of climatologists have compiled detailed data on temperature and rainfall in Mongolia based on tree-ring growth, making it possible to determine the self-calibrating Palmer Drought Severity Index (scPDSI), a measure of aggregate water balance, for the past 1,112 years.[17] Their findings were:

1115–1139 CE and 1180–1190 CE were extremely dry, with periods of below-average reconstructed scPDSI. The 1180s drought occurred during the turbulent early years of Chinggis Khan, a historical period characterized by warring tribes and factions on the Mongolian steppe (18). These droughts were followed by a wet period beginning in the early 13th century and then a return to drier conditions until the late 14th century. From there, multiannual to decadal-scale variation in hydroclimate continued until the 20th century when climate was wetter than any other century since 900 CE.

Moreover, for the specific years when Chinggis Kahn was consolidating his empire:

> The most unusual period in the 1,112-y record was not marked by extreme variability but was instead persistently wet. From 1211 to 1225 CE, no annual values or their bootstrapped confidence limits drop below the long-term mean of the reconstruction, making it an unmatched "pluvial," a prolonged period of above average moisture, over the last 1,112 y.[18]

The authors of the study point out that the dry climatic conditions of the late-twelfth century coincide with a period of extreme political instability in Mongolia, which culminated in the rise of Chinggis Khan, while the period following his seizure of power correlate with a dramatic increase in precipitation that greatly increased the availability of energy, in the form of grass and thus herds, in the Steppe, providing resources to strengthen his rule and making possible larger concentrations of horses and flocks and thus of people and armies.

The authors of the study are careful to avoid saying that climate change "caused" the expansion of the Mongol Empire. Their vocabulary, however, much as that pointed out by Hempel, that climate was a "factor," that climate "promoted" high grassland productivity and "favored" the formation of Mongol political and military power, strongly suggest a causal connection between the climatic conditions they document and the rise of the Mongols.

Could one deduce from this study a "law of history," and, if so, how broad or restrictive would it be? Remember that to be a law, if one postulates that a series of specific determining factors resulted in an event of a certain kind, then the implication is that the presence of these factors will necessarily produce the same outcome in the future. Might one postulate: "Empires rise and fall with climatic change"? This would be absurd: the Mongol Empire was a very special type of empire arising in a very unusual environment particularly sensitive to variations of drought and rain. One could hardly postulate that other empires, be they the Roman Empire, the Persian, the Ottoman, or the British, all necessarily responded similarly to similar climatic changes.

Could one argue that "Drought causes political instability and conflict, while generous rainfall increases economic and political power"? This might appear common sense, but it too is dubious: A law should have predictive power or at least it should be capable of showing replication. But in other contexts, the Byzantine Empire of the sixth through twelfth centuries, for example, although paleo-climatologists can track significant variations in rainfall across Anatolia (modern eastern Turkey), in fact, one finds political, social and economic stability in spite of such fluctuations.[19]

Perhaps the "law" should be formulated in an even more restrictive manner: "In pastoral economies, periods of drought destabilize polities and periods of abundant rainfall increase the economic and political power of such polities." This may be closer to the mark, but we have moved to a level of generalization that is not particularly helpful. For one thing, we have lost the concept of empire, let alone the formation of the greatest empire in world history, and its applicability is now only to a specific and rather limited type of traditional economic and political regime. We do, however, have something of a confirmation of this "law," at least in Mongolia: The same team found that wet conditions during the twentieth century in Mongolia corresponded to a period of widespread agricultural development, while a period of severe drought between 1998 and 2010 resulted in the death of millions of animals and, in the period of 1999–2002 alone, the migration of approximately 180,000 people or 7.5% of the total population of Mongolia from the steppe to the capital city of Ulaanbaatar.[20] But how useful is this as a "law of history"? After all, the period of high grass production did not produce a Chinggis Khan or another Mongol Empire, nor did the period of drought see the collapse of the Mongolian state.

Is the relationship between climate and the rise and fall of polities not so much a "law" as an example of Hempel's explanation sketch, that is, "a vague indication of the laws and initial conditions considered as relevant"? And would one need to examine other "conditions" or "causes," including climate, means of production, technologies, and ideologies not only with Mongolia of the thirteenth century, but in China, Western Asia, and elsewhere before completing Hempel's formulation C_1, C_2, ... C_n produce E. If so, then unless one could find exactly the same series of C_1, C_2, ... C_n, something that I would contend is impossible, the formulation of such a law is meaningless. It is little more than Toynbee's flaccid challenge and response. Thus, wisely, Di Cosmo and his colleagues make no claim about the laws of history or about the necessary effects of drought and rainfall on the rise and fall of empires. They are content to show a powerful correlation between climatic change and the rise of the greatest empire in world history: This, I think is sufficient.

The difficulties of postulating a "law of history" even in the case of a society with a simple, direct relationship to a single, overriding factor such as rainfall, should be clear. The problem becomes even more complex if one looks at more complex historical debates such as that which has come to be called "the great divergence." How, historians want to know, did Western Europe and parts of North America become the dominant political and economic nations of the world? This is hardly as simple a question as "did environmental change bring about the rise of the Mongol Empire?" but it is a fundamental one. In particular, the question is not only when did the economic and political power of Western Europe eclipse that of the other great polities such as India, China, the Ottoman Empire, and Japan, but what were the root causes of this divergence. Within the context of this workshop, the question might be rephrased as follows: "Were there specific causes for the Western emergence as the world power, and, if so, when and where did these first appear? And if there were indeed 'causes,' can we then postulate the laws that explain why these causes led to Western dominance?"

Traditionally, European and North American historians have posited geography, natural resources, colonialism, customs, social and economic traditions, or some mix of all of these to explain the rise of Western dominance. The time when these factors is said to have begun has been variously placed in the fifteenth and sixteenth centuries — the "age of discovery" or the Renaissance; the rise of European colonial empires or the scientific revolution in the seventeenth century, or even the Enlightenment of the eighteenth. European "exceptionalism," was understood to be part of the history of Europe from even earlier, perhaps as early as the commercial revolution of the twelfth century. These have all been posited as necessary precursors to the industrial revolution of the late-eighteenth and early nineteenth centuries, which made the economic development of Western Europe uniquely possible.

Most recently and more fruitfully, debate has focused specifically on those factors (causes?) that resulted in the economic capacity of the West to exceed those of other contemporary polities. Why was Europe able to break out of the Malthusian cycle of stagnation and dominate the world's economy in the nineteenth and twentieth centuries? Such questions, while academic, are also potentially significant in terms of understanding future global change. If one can isolate the "laws" behind the great divergence, can these same conditions be replicated for other polities attempting to achieve rapid economic growth and with it an enhanced place in the world? Also, if one can pinpoint the causes of the great divergence, can one then predict what some are terming the great convergence, the rise in economic productivity in other regions of the world that increasingly challenge Western dominance?

The most impressive recent attempt to address these questions is the 2000 book by Ken Pomeranz that offers a challenge to traditional explanations for the economic dominance of the West in recent history.[21] Pomeranz reviews the various factors that historians have attempted to argue as causative of the European takeoff, but he does so by a fundamental recontextualization. Rather than treating the conditions of Europe as exceptional, he undertakes a strict comparison of such factors as agricultural, commercial, and proto-industrial development among various parts of Eurasia as late as 1750. Likewise, he eschews such generalities as "Europe" and "Asia," comparing instead smaller regions such as England and the Lower Yangzi Delta, that is, those areas that were most similar in all of these factors. This allows him to consider those areas of the world most similar in such things as population growth, income, manufacture, agrarian practice, labor systems, markets, institutions. Through this comparison, older theories such as differences in class struggle, labor relations, industrial productivity, and the like can be eliminated as significant differences. To return once more to Hempel's formulation C_1, C_2, ... C_n produce E, one could say that Pomeranz shows that these Cs are present in significant areas of both England and China, and yet in one region E is produced; in the other it is not. In fact, he demonstrates that the Lower Yangzi Delta in 1750 was by most of these measures in advance of England. Postulating these factors as causes, much less arguing that some abstract laws, whether economic, social, or institutional, were at work to produce the European miracle, (which of course was not European and certainly no miracle) simply doesn't work.

In his systematic comparisons of his select regions in Europe and Asia in terms of consumption, investment, capitalism, as well as ecological factors and agricultural technologies, he effectively undermines any argument that some abstract laws dictated that Britain rather than some region of Asia would be the center of economic progress in the nineteenth century. He ultimately focuses on two advantages of England that, in his words, "were not advantages that had to lead to an industrial breakthrough, but advantages that greatly increased that possibility and made such a breakthrough much easier to sustain."[22] These advantages might be summed up in two words: Coal and Colonies. First, the largely fortuitous discovery of large coal deposits and their exploitation gave England a source of energy that allowed it to overcome the environmental constraints imposed by limited supplies of wood for fuel. China, of course, also has vast stores of coal that were already being exploited in the eleventh century. They are in the north, however, far from the most developed areas of China, and exploiting them raised very different challenges to those of English coal mines, and the industry seems never to have recovered from the devastation brought about by the Mongols in the thirteenth century. By chance, English coal mines were close to the most advanced areas of economic development and thus their exploitation not as great a challenge. England, he argues, was simply lucky in its coal.

The second enormous advantage to England and some other parts of Europe was access to the resources of Africa — slave labor — and the Americas, chiefly precious metals which could be used to acquire Chinese and Indian goods and increasing amounts of land-intensive food, fiber, and timber. Access to New World resources, aided by the vulnerability of Amerindians' vulnerability to smallpox, fortuitously allowed the most advanced regions of Europe to escape the ecological and demographic constraints that limited other regions. These two fortuitous factors, that the English had the right coal in the right place and that Europeans obtained virtually unopposed access to American resources, are at the heart of Pomeranz's explanation of the great divergence, the take-off, first of England and the Low Countries, and then much of Western Europe and its North American offspring. He is quick, however, to combine these fortuitous events with what he calls other conjunctures, such as the Chinese demand for silver, favorable wind patterns for trans-oceanic shipping, and disease gradients, all of which allowed parts of Europe to diverge from what had been, well into the eighteenth century, more prosperous and highly developed Asian societies.

Assuming for the moment that Pomeranz is correct (and of course not everyone would agree with him)[23], can one begin to talk about laws that govern economic development in the past or that might predict it in the future? How would such a law be formulated? "Advanced economies that happen to stumble across an easily exploited new source of energy and an easily exploited new world are most likely to achieve an industrial revolution?" This would hardly be a meaningful norm: it is certainly not repeatable. Perhaps we would need to amend the Hempel formula thusly: $(C_1, C_2, \ldots C_n)$ when accompanied by precondition $(P_1, P_2, \ldots P_n)$ will produce E. However I see little value in such a formulation. At best this is an explanation sketch, which suggests the complex of conditions that, at one time, produced a unique result. However these conditions will never be exactly repeated, any more than will the conditions that produced the Mongol Empire return.

Of course, a physicist might object: According to the Poincaré recurrence theorem, any finite system will eventually return arbitrarily close to its initial condition.[24] Thus, eventually, exactly the same conditions and situation of Mongolia in the thirteenth century will return. We can then verify if we have identified the laws that underlay this event. However, since the time it would take for such an event to occur can be estimated to be at least $10^{10^{19}}$ times the lifetime of the universe, as an elderly historian I rather doubt that I will live to see this happen.

However, even if the conditions that produced the Mongol Empire or the Great Divergence are never precisely repeated, understanding the particulars of these conditions is, I believe, important and worthwhile. It may even help us to understand how similar, but not identical conditions may produce certain future results, but these results will undoubtedly different from those of the past. The Greek Philosopher Heraclitus had it right: in history, unlike in physics, "You could not step twice into the same river,"[25] or as the American humorist Mark Twain is said to have opined, "History never repeats itself but it rhymes."

References

1. S. Weinberg, *Dreams of a Final Theory: The Scientist's Search for the Ultimate Laws of Nature* (Pantheon Books, 1992), p. 35.
2. *Ibid.*, p. 36.
3. *Ibid.*, p. 39.
4. D. A. Duquette, "Hegel: Social and Political Thought," *Internet Encyclopedia of Philosophy,* https://www.iep.utm.edu/hegelsoc/
5. https://www.iep.utm.edu/hegelsoc/
6. P. Blackledge, "Historical Materialism" in *The Oxford Handbook of Karl Marx,* eds. M. Vidal, T. Smith, T. Rotta, and P. Prew (Oxford University Press), DOI: 10.1093/oxfordhb/9780190695545.013.1.
7. A. Toynbee, *A Study of History*, 12 Vols. (Oxford University Press, 1948–1961).
8. A. J. Toynbee, "The Challenge Hypothesis," http://www.panarchy.org/toynbee/challenge.html.
9. D. Hume, *A Treatise of Human Nature* (London: 1739), T 1.3.11.4 (http://www.davidhume.org/texts/thn.html)
10. K. G. Hempel, "The Function of General Laws in History," *J. Philosophy* **39**(2), 35–48 (1942).
11. Hempel, p. 35
12. Hempel, p. 35.

13. Hempel, p. 36.
14. Hempel, p. 36.
15. Hempel, p. 40.
16. Hempel, p. 42.
17. N. Pederson, A. E. Hessl, N. Baatarbileg, K. J. Anchukaitis, and N. Di Cosmo. "Pluvials, droughts, the Mongol Empire, and modern Mongolia," in *Proc. Natl. Acad. Sci. U.S.A.* **111**(12), 4375–4379 (2014), doi.org/10.1073/pnas.1318677111.
18. Pederson *et al.*, p. 4376.
19. A. Izdebski, J. Pickett, N. Roberts, and T. Waliszewski, "The environmental, archaeological and historical evidence for regional climatic changes and their societal impacts in the Eastern Mediterranean in Late Antiquity," *Quat. Sci. Rev.* **136** (15 March 2016), 189–208, doi.org/10.1016/j.quascirev.2015.07.022.
20. Pederson *et al.*, p. 4378.
21. K. Pomeranz, *The Great divergence: China, Europe, and the Making of the Modern World Economy* (Princeton University Press, 2000).
22. Pomeranz, p. 211.
23. See for example the long review by P. H. H. Vries in *J. World History* **12**(2), 407–446 (Fall, 2001).
24. J. Galkowski, "A Survey of Results in Poincaré Recurrence Estimation," https://web.stanford.edu/~jeffg1/Papers/estimationofpoincarerecurrence.pdf.
25. δὶς ἐς τὸν αὐτὸν ποταμὸν οὐκ ἂν ἐμβαίης, quoted by Plato in his *Cratylus* 402a.

Chapter 3

The Nature of Laws and Principles in Science

David Gross

Kavli Institute for Theoretical Physics
University of California, Santa Barbara, CA

Introduction

The central theme of this meeting is "Laws, Rigidity and Dynamics". I am not exactly sure what rigidity and dynamics mean, or what "laws" refer to, as we have learned that the meaning of laws differs so much across all of the various fields represented here. I will take a strong (somewhat arrogant) scientific point of view, and discuss what role "laws" play in Science,

I will begin by discussing the Scientific Method, and then try to define what I mean by facts, truth, theories, principles and laws, as these are produced by the Scientific Method. I will discuss the role of observation, experimentation, models and theories.

Laws, theories and explanations in the Natural Sciences, and especially in physics, are always formulated in mathematical terms. This is one of the reasons that these multidisciplinary exchanges are so difficult. To some extent, how can one explain the grandeur of Beethoven's music to someone who is deaf? (Beethoven was also deaf, but he could somehow appreciate his own music, perhaps because he was able to hear for most of his life.) To the innumerate, it is hard to convey what the laws of physics really are. Mathematics is essential, so I will discuss the role of mathematics. Then I will give some examples of laws and principles in physics, and discuss the reductionist nature of physics.

I will end by asking whether we could have a final set of laws, a final theory, what that might mean and if there is a final theory can we find it?

Truth

A few months ago, I had the opportunity to talk about Truth at a very interesting event — the Nobel Dialogues that were held in Gothenburg on the day before they gave out the Nobel Prizes. This year, the Dialogues were devoted to the "Future of Truth" — you might guess why. Among the participants were scientists, but others as well, such as Maggie Haberman (the White House correspondent for the New York Times) and the ex-Director of the CIA. They reported some very interesting and very depressing news regarding the status of truth in the USA.

I gave one of the opening talks, on "Truth and the Scientific Method." Preparing for that talk, I tried to understand what "Truth" is. I looked in many dictionaries and discovered that nobody actually defines Truth. These sources describe various properties of Truth, but never Truth itself. Most definitions of truth are circular, essentially stating: Truth is what is true! For example:

Truth is a quality or state of being true. Merriam-Webster
Truth: that which is true, or in accordance with fact or reality. Oxford English Dictionary
Truth: agreeing with fact, not false or wrong. Cambridge Dictionary

I decided the best definition I could think of for Truth is: **Truth is that which is revealed by the Scientific Method**. Thus, I am replacing Truth, (something that humans once thought was absolute and real and determined by a deity or his representative on Earth), by a process. Which process? The process of the Scientific Method. I believe that what the scientific method reveals is the closest candidate that we have for Truth. As we shall discuss, this Truth is expressed in regularities, laws, and theories — but it is not absolute, rather it is always tentative and subject to revision.

The Scientific Method?

The Scientific Method rests on assumptions that conflict with much of what was believed for most of human history. It rests on the assumption that the world of observable phenomena is real and intelligible (in a collective manner, rather than an individual manner), that our understanding of nature is based on observation and experimentation, and is subject to the requirements of logic and consistency. No theory or mathematical argument, no matter how compelling or beautiful, can be maintained in the face of contradiction with observation or with experiment. Nowadays, these are accepted principles that are taught in elementary school (unfortunately, most of the population soon forgets them) but they were revolutionary when the Scientific Method was discovered and advanced by humankind four hundred years ago. Over the last few hundred years these principles have been perfected and continue to be perfected. I regard the Scientific Method as the greatest discovery of our species.

In very broad terms, there are three essential components of the Scientific Method:

Observation, Experiment and Theory

Observation and experimentation are quite different. Most sciences begin with observation. Some are still stuck in the observational phase. In some sense, observation addresses 'what' questions, such as "What is that newly observed phenomenon". Experiment is more elaborate than observation. Experiments are more informative and precise ways of probing observed phenomena. Experimenters set up controlled experiments and perform measurements that can often be expressed in quantitative, precise, mathematical form and address 'how' questions, such as "How does nature work?" The ultimate goal of the Scientific Method is to produce an explanation or theory ("Truth") which can describe in quantitative form "how things work", and also often addresses the question of "why."

The Scientific Method consists of a very tight relation between theory, experiment and observation. Theory informs experimenters or observers what to look for and suggests which precise measurements to perform. Experiment puts theories to the test. Observations and experiments discover new phenomena. And finally, both the experimental understanding of how things work, and the theoretical ability to predict how things will work, allows one to apply scientific truth to the real world to create new tools and control our environment. The application of scientific truths to develop new technology is an essential part of the Scientific Method. Its success in increasing our control over nature and in developing new technologies explains why society supports science and why society (often reluctantly) accepts its findings and is willing to overturn previous superstitions, myths and other non-scientific explanations of natural phenomena. Also, new technological tools feedback directly into improving our ability to observe new phenomena with ever increasing precision, often by orders of magnitude every decade or two.

It is this Scientific Method that produces the best candidates we have for Truth. What is special about scientific Truth, the product of the Scientific Method, is that its findings, facts and theories, are subject to the requirements of logic, consistency and continual confrontation with reproducible experiments. Scientific truths are always tentative. Consequently, given my definition of Truth, there is no absolute Truth which we can discover or create — scientific Truth is always tentative. It always must subject its tenets to experimental test.

In the end, the only judge of scientific Truth is Nature, and Nature can never prove that our Truths are absolutely true. However, Nature can prove that a theory is not true, by performing experiments that are in conflict with the predictions of a theory. As Feynman eloquently said, "we are never definitely right — we can only be sure that we are wrong." We subject our Truths to Nature's tests, and often nature says, "It's okay; in this case, I agree." That does not mean that the theory is true, as in another case, a different or more precise experiment, nature will say, "I disagree," and then you must alter or discard the theory. Nature is a very strict judge. Thus, subjecting one's ideas to Nature's judgement can be very frustrating. In the end, what convinces us of the validity of a scientific theory is that it works, over and over again. If you ask a scientist why they have a (tentative) belief in their theories, it is because these theories have worked over and over again, which means the predictions of the theory have not been disproved over and over again. At some point the theory becomes accepted Truth.

There are certain aspects of the nature of scientific truth that are very important, and that have not necessarily been widely adopted outside of science.

1) **Science is liberated from its creators**. It belongs to no one and it belongs to everyone. The scientific culture that grew around the Scientific Method has gone to great pains to assure that people cannot copyright or own scientific Truth. The results and implications of science are available and relevant for everyone.

2) **Science is universal**. It applies everywhere and to everyone. Unlike many other norms, regulations, laws, culture, and beliefs, the laws of science and the facts and the truths revealed by science are universal.

3) **Science lacks all authority except Nature, and all are equal in its pursuit**. Of course, scientists are humans, and they work in human institutions, with all their shortcomings. But, in the end, it is the case that all are equal in the pursuit of scientific truth, in the sense that any valid claim that is not contradicted, or predictions that are not disproved, will, in the end, win out, as there is no other authority except for nature.

Scientific truths, unlike many other human beliefs, are tentative, always subject to change and indeed are continuously changing. The essential part of the Scientific Method is the necessity of subjecting one's predictions to experimental tests. These tests can only falsify scientific theory, they can never prove. One cannot prove that a theory is correct in science and physics, you can only disprove and falsify it.

The Scientific Method has been remarkably powerful and successful in understanding and controlling nature, but it has also had a great influence on society. It is not an accident that the enlightenment and the beginnings of political democracy followed the evolution of the scientific method. Although science is not a democracy and the laws of nature are not adopted by majority vote, but by correspondence with nature — yet everyone has an equal say about what are the possible laws and theories. The characteristic features of the Scientific Method of openness, inclusiveness and transparency, that are necessary for it to work so well, promote these same values in society.

The Role of Theory

Most fields of science start out being rather observational and experimental, but the most advanced sciences — such as physics — have become highly theoretical. Currently, within physics (and, more and more, within chemistry, biology and ecology as well), theory is assuming a more important role. This was not always the case; theoretical physics wasn't recognized as a separate field until the mid 19th century. But now, within physics, there is a sub-community of people who do not engage in observation, who do not perform experiments, but rely on the findings of those that do to develop and test the laws and theories. The role of these theorists is to:

1) Quantify and model observations.

Originally, theorists mostly quantified and modeled observation. That's still how many make their living, fitting data with existing explanations (theories and models), and often discovering new patterns and regularities in the data, an extremely important service to experimenters and observers.

2) Suggest new experiments or observations, and specify the required necessary precision to test these ideas.

Tentative explanations, theories, and quantitative models can suggest new experiments and observations, thus giving the experimenters and observers hints as to where to look and what discoveries might be made. But most important, and increasingly so as the science becomes more and more mature, specifying the necessary precision to subject theories to falsification. You must have a theory in order to be able to falsify something.

3) Predict new phenomena, or predict the precise value of measurements, so that these predictions can be used to falsify general principles or specific theories.

With enough non-falsifications, we develop trust in the theory — not proof but trust.

4) Model and calculate the expected background to new experiments, so that experimenters can discover new phenomena and distinguish signals from noise.

Astronomy functioned for years as an observational science. One looks up and sees the stars — although they are pretty, you don't learn much. But, once you had a theory of atomic and nuclear structure, you could analyze light from all those stars and discover new phenomena. The theorists could then tell the experimenters what telescopes and instruments they needed in order to understand the structure of the universe, its history, and to distinguish signals from noise. All data is noisy, and the understanding of the data itself requires lots of theory. One of the important roles of theory is to help experimenters perfect their measurements.

5)   Finally, the ultimate goal of theory is to accomplish the reductionist goal of science, that of unifying disparate phenomena and reducing complex phenomena to simple constituents and basic laws of dynamics.

Mathematics

As physics and science matured, we learned that we need to follow Galileo's dictum, and use mathematics to formulate the laws of Nature. Galileo described beautifully that the language of Nature is mathematics:

Philosophy (by which he means physics) is written in that great book, which ever lies before our eyes — I mean the universe — but we cannot understand if we do not first learn the language and the grasp the symbols, in which it is written. This book is written in the mathematical language, and the symbols are triangles, circles, and other geometrical figures, without whose help it is impossible to comprehend a single word of it; without which one wanders in vain through a dark labyrinth.

The necessity of mathematics has been forced on us by the structure of nature. Over the centuries, we have discovered that as we probe deeper and deeper into the structure of nature more complex and abstract mathematics is required, and that there is an incredible congruence between the language of nature and the language of mathematics. The fact that we use quantitative mathematical analysis in formulating the laws of physics is partly what enables us to achieve incredible precision in our experiments and in the tests of our ideas.

I cannot resist discussing a famous example of precision in experiment and theory. I refer to the so-called *g*-factor: the pure number that measures in fundamental units the magnetic field of a single electron. An electron is a spinning charged particle. Because it is spinning, it creates a magnetic field, a little magnet whose strength is characterized by the number *g*. In the theory of Quantum Electrodynamics, the theory of electrons interacting with light, g is approximately equal to two. The difference, $g - 2$, can be calculated, with an enormous amount of effort, using the theory of Quantum Electrodynamics (QED), and some of the other parts of the standard model of particle physics, to one part in a trillion:

$$(g-2)/2_{\text{theory}} = 0.001\ 159\ 652\ 181\ 643(764) \tag{1}$$

Experiments to measure this number, with this precision, are incredibly difficult, requiring the observation of a single electron, isolated from the environment, and trapped in a magnetic trap. These beautiful and amazing experiments have measured g-2 to one part in a trillion as well:

$$(g-2)/2_{\text{experiment}} = 0.001\ 159\ 652\ 180\ 73(28) \tag{2}$$

The amazing thing is that these numbers agree to one part per trillion, up to the expected errors in both theory and experiment. Physicists are extremely proud of both our experimental prowess and the standard model of particle physics, wherein we can measure and can calculate up to one part in a trillion. Caring so much about precision is part of the tradition and culture of physics. But why do we expend all this effort? I mean, if QED works to one part in a million, isn't that good enough? We care about these statements, these laws and these theories, and continually test them. The comparison of theory and experiment to one part in a trillion continues, because we expect that, at some point, every theory will fail. And that failure will teach us something. Science is continually looking to falsify itself, in order to learn more.

To have this kind of agreement of a highly mathematical law with Nature means, as Galileo suggested, that there is something inherently mathematical about Nature. Mathematics appears to describe the real world very well — it is the natural language, and the appropriate language to describe Nature. Some have regarded that as strange. Wigner wrote of the "unreasonable effectiveness of mathematics". I do not see it as strange, because I think that both mathematics and physics are products of the same evolutionary process. Consider space and time. At a very early age each of us construct a model of space and time. This is just a model, as space and time are not things we directly feel or see. Space and time are mental constructs that we invent to understand and control the physical world. We all went through this in our first year while our brain was growing and developing. Each of us, without any instruction, constructed a model of space and time. I regard it as perhaps the greatest intellectual achievement that each of us achieve in our life.

The model we construct is roughly three-dimensional Euclidean space, a pretty good model for the non-relativistic, low energy world. But, as we learned in the 20-th century, it is actually wrong in many respects and fails when objects are moving fast or gravity is taken into account. Our minds are shaped by evolution to construct tools to understand and control the natural world. The model of space and time was driven by the need to be able to cross the room, and get the toy. Our species is driven to understand and control the world, and to construct the tools that enable us to do that, mathematics being one of them. It is therefore no surprise that the tools we construct to understand and control the physical world are so effective. That's why I believe that mathematics works so well in describing the physical world. It would be unreasonable if it were not so effective.

Reductionism

The strength of science is greatly amplified by its reductionist nature. Philip Anderson said that "science is a multiply connected web", a feature that is a source of great strength. Science does not consist of individually constructed explanations, one explanation for this, and another explanation for that — it is all connected — and that is a source of great stability and power.

Reductionism is what gives science its structure. We have learned that complicated things are made out of simpler things, which in turn are themselves made out of simpler things. These are simpler in the sense of smaller constituents, but not necessarily in the sense of the nature of the highly mathematical theories we must construct as we go to a smaller and smaller distances, until we are farther and farther removed from our microscopic experience, and we need to modify our laws and explanations in ways that are very difficult conceptually, and which require very deep, new mathematical tools.

A strong reductionist point of view would say, that in principle, one can start at the bottom with our best theory of Nature at the microscopic level and, in principle, deduce our current understanding of the laws of gravity, of nuclear force, of electromagnetism, atomic structure, and nuclear structure. Everything follows in this reductionist sense. We indeed already have an incredibly successful theory of all the forces that act within the atom and the nucleus, as well a classical theory of gravity. All macroscopic phenomena, in principle, can be deduced by mathematical means from this microscopic theory, if you had the time, the computational power and the will. In most cases, of course, this would be a hopeless and useless task. Instead we use shortcuts and approximations. However, it is a central feature of nature, that all that occurs at large distances, and in more complex collections of the basic constituents of matter, could be deduced from the laws fundamental physics that govern the microscopic constituents of matter. This includes Biology. Therefore, life itself, our brain, our norms, everything — is ultimately explainable in terms of fundamental physics. This reductionism has been extraordinarily successful. We have not found any counterexamples, although many over the centuries have been postulated, especially in the case of life and the mind.

This connected reductionist structure of nature gives the scientific truths revealed by the scientific method and its laws enormous power. Thus, one cannot argue that CO_2 emissions do not cause global warming without contradicting the truth of physical chemistry, and thus truth of the quantum theory of atoms and the whole web of fundamental physics. The behavior of CO_2 in the atmosphere is, most scientists agree, not just an assumption of climate scientists, but it is a consequence of what we understand of the chemistry, and chemistry is a consequence of what we understand about the atom and quantum theory, and so on. If you try to remove one brick of this immense structure, the whole thing collapses. Consequently, scientific Truth is solid, in a way that many other candidates for Truth are not; solid, if not always convenient. That is not to say that everyone agrees with this. They will argue with the observations, or invent elaborate explanations of why the arguments fail. So, I am not arguing that there is any way of proving scientific truths. As I've said before, you can never prove scientific truths, you can only disprove.

The Limitations of Science

I have discussed the Scientific Method which produces science, which I regard as equivalent to Truth, but it has limitations. However, there are limitations to this method, and to the Truth that it produces, as there are questions that, currently, cannot be addressed by observation, experiment, or mathematical theory. Often, when I give a talk, I get questions from the audience, "So you have this incredible theory, tell me "Why there is something rather than nothing?"; or "What is the meaning of life?" These are interesting questions, which humanity has been asking (for eons), but they cannot, at the moment, be addressed by the Scientific Method. So even though I myself ask these questions, and would be interested in their answers, I'm not that interested because I can't address them scientifically, and therefore have little faith that they will ever be answered. Obviously, the Scientific Method cannot address questions that cannot be addressed by the Scientific Method. But it sometimes happens that questions that were once thought to be outside the domain of science, such as "What is Life?" and "What is the history of the universe?", and previously left to religion and philosophy, are now part of science and addressable by the Scientific Method. We certainly have a good theory of what life is, and how it originated. We now have a theory of the history of the universe, based on observation, and the theory works pretty well, and could be falsified, and is continually subjected to tests. Remarkably, physical cosmology is now seeking to address questions like, "How did the universe begin?" which is now becoming a scientific question, addressable by the Scientific Method. Will the question "Why there is something rather than nothing" be addressable someday by the Scientific Method? Perhaps.

Laws in Physics

The terminology regarding 'laws' in physics is random and often confusing. There are laws, there are principles, and it is very hard to distinguish which part of a scientific theory is a law, a principle or simply an equation.

I couldn't really figure out what's the difference between a Law, a Theory, a Principle, a statement, a theorem or an equation. These terms are used interchangeably, and which is used, is often an historical accident.

The best example of laws of nature are Newton's laws of motion, the first mathematically comprehensive theory of Nature which sparked the initial explosion of the scientific revolution. Newton's laws of motion relate an object's motion to the forces acting on it. The first law states that an object will not change its motion (its velocity) unless a force acts on it. The second law states that the force on an object is equal to its mass times its acceleration. In the third law, when two objects interact, they apply forces to each other of equal magnitude and opposite direction. The first and third law are actually consequences of the principle of Relativity, a symmetry principle, initially formulated by Galileo, that states that any observer moving with a constant velocity will experience the same physics. The principle of Relativity is a very deep principle, a principle of symmetry, which essentially says there is no "privileged observer". It is the basis of two Newton's Laws, and the basis of Einstein's theory of special relativity, and eventually the general theory of relativity which explains gravity. The symmetry principles, not laws, lead to conservation laws which explain why energy and momentum are conserved. The second law is a definition of inertial mass and some properties of motion, namely that a force on a body causes acceleration (change of velocity). However, unless the force is specified, this law is somewhat empty. In the case of gravity, the force was specified by Newton's law of gravity, which states that the force of gravity between two bodies is proportional to the product of their masses and inversely proportional to the distance between them.

Newton's framework of physics and his laws of motion and of gravity dominated physics until the 20th century, until the revolutions of quantum mechanics and special and general relativity. So, Newton's laws are wrong. They've been overthrown! Paradigm shift! They have been replaced by Quantum Mechanics and Einstein's theory of general relativity. But this is misleading, Newton's Laws are not wrong and they certainly have not been replaced. If one wants to go to the moon, one doesn't use Einstein's theory of relativity. Rather one uses Newton's laws — they are fine as the corrections dictated by Einstein's theory are negligible in this case. Newton's theory is not overthrown or discarded. Rather it survives as a well-defined, limiting case of Einstein's theory. This is true of all successful theories in physics. Inevitably they have been, or will be, shown to be, in some sense, wrong. But in another sense, they're not overthrown or discarded, nor will they ever be. They survive as well-defined limiting cases of larger, more comprehensive, more precise, "truer" theories.

Some other laws in physics laws are kind of trivial, such as Ohms law, that describes electrical circuits. Ohms law tells you that if you put a voltage across a circuit conductor, a bunch of metal, that the current flowing is linearly dependent on the voltage, and that the coefficient is called the resistance of the circuit. One of my mentors told me that in the US Army, where is he served in the Signal Corps, he was taught by his instructor that there three Ohms law. There is: $V = I \cdot R$ or $I = V/R$, $R = V/I$. I'm sure you can understand, without too much mathematics, that there is really only one Ohms law — which isn't much of a law anyway, and it's not always exactly true, as it's a kind of an expression of regularity, a useful 'law' to teach people who need to fix radios.

Then there are laws that are really important, such as the second law of thermodynamics. This law, as Einstein explained, is probably the only law in physics that will survive forever. Einstein believed, as we all do, that truths, explanations, theories, and laws in physics are always tentative, and we are pretty sure that they will continually be replaced by something better. They will be disproved at some point, in some form, and then contained in some larger explanation. With the one exception, Einstein argued, namely the second law of Thermodynamics, which should survive forever. He did not believe that his own laws, (which aren't called Einstein's laws, they're called Einstein's equations), would survive forever. But the second law of Thermodynamics is a different kind of law, and, more importantly, is the kind of law that is much more likely to have applications in other fields, such as social sciences, because it's simply a consequence of thinking about the behavior of complex systems, which have many component parts and many constituents. It is a consequence of the laws of probability and the law of large numbers, and not specific dynamical theories.

Then there are principles, such as Archimedes' principle: The upward buoyant force that is exerted on a body immersed in a fluid is equal to the weight of the fluid that the body displaces. This principle can be derived from our understanding of fluids, using Newton's laws, or Quantum Mechanics. It is a consequence of our theories and the laws they contain, so we shouldn't call it a principle anymore. But back then, in in the absence of a comprehensive theory of fluids it was called a principle and was very useful for measuring volume. Also, it is not exact, as it neglects surface tension and it fails for complex fluids (such as shaving cream.) To derive it you must make certain assumptions, such as neglecting surface tension.

However, there are principles which are really sacred and very important; for example, Heisenberg's uncertainty principle, which captures an essential element of quantum mechanics. I have no idea why is it called a principle and not a law.

Anyway, my impression about the terminology — laws, principles, equations, etc. — that we use to label the structure of physical theories and explanations, is very loose and arbitrary.

Frameworks, Theories and Models

Currently, our best reductionist theory of nature is called the Standard Model. It describes, we believe, with only a few parameters, almost everything we have ever observed, in a reductionist sense. The standard model is truly a theory, not a model. The structure of physics is that we have frameworks, theories and models and one should distinguish these.

A framework is an overall structure of theories. For example, quantum mechanics is currently one of the most important aspects of the current framework of physics. Quantum Mechanics is not a theory. A theory is a specific set of laws within the framework of physics. A theory is something that can be falsified. To falsify quantum mechanics, you actually need to have a specific theory of something consistent with the rules of quantum mechanics, that can be tested in a laboratory. Quantum Mechanics is a framework in which many possible theories reside. Similarly, Quantum field theory, which is the basis of the Standard Model, is also a framework.

The Standard Model is a theory, consisting of a set of laws within the framework of quantum field theory and a set of parameters that must be determined by measurement. In this theory, one can in principle calculate, as we've been doing for the last forty years, the results of thousands of experiments with increasing precision, anywhere from one part in a hundred to one part in a trillion. That is a theory.

Non-relativistic quantum mechanics, or the quantum mechanics of atoms, is the most successful theory we've ever had. It actually describes everything you see around you in this room: the behavior of atoms, collections of atoms, solids, gases, proteins, light, and us. Basically, it has one parameter in it, which measures the strength of the electromagnetic interaction. So, in a reductionist sense, in the 20th century, we ended up with, an "in principle" theory for just about everything that concerns us macroscopically, (with just one parameter). This theory doesn't account for the inside of nuclei, which is why we need the Standard Model (more precisely the standard theory).

Finally, there are things that are called models. For example, the so called BCS theory of superconductivity is actually a model. It is a simplification, within the quantum mechanical theory of atoms; a drastic simplification which, however, is a good enough approximation or simplification to account for the remarkable phenomena of superconductivity. In fact, it fails as a model for the phenomenon of high temperature superconductivity, it has been disproved. Now that does not contradict quantum mechanical theory of atoms, or the whole framework of quantum mechanics, it just says that some of the simplifying assumptions in the BCS model were wrong, in the case of high temperature superconductivity.

Most of what is talked about, in other fields of science, are models but not theories. And when they're disproven, and they should always be subjected to being disproven, then you must change the parameters, or expand the model to include phenomena that you thought you could ignore before. So, it is very important to distinguish these levels of understanding of laws or explanations of theory. For example, consider one of the frontiers of current research — string theory. String theory originally was thought to be a revolutionary break from quantum field theory, the current theoretical framework of fundamental physics, as it replaces point-like elementary particles with extended strings. What is so exciting about string theory is that it automatically contains gravity in addition to all the other forces of the standard model. However, critics of string theory complain that string theory is unscientific as it is unable to make predictions that could be used to test or falsify the theory, without doing experiments at extraordinarily small distances or high energies, way out of current experimental reach.

This criticism is correct, not only because the scale of the new physics envisaged by string theory is so far removed from current observation, but also because string theory is not really a theory, but rather a framework. Much like quantum field theory, which is really a framework which contains many possible specific quantum field theories, such as the standard "model", there are many possible string "theories' and we have neither the experimental clues nor the principle that would choose which one of these (if any) describes the real world. Furthermore, in recent years we have realized that these two frameworks — quantum field theory and string theory — are connected, and can often describe equivalent, dual, descriptions of the same phenomena. These frameworks are thus part of some greater framework whose full scope we are currently exploring.

A Final Theory

I have discussed the scientific method and its accomplishments. Any discussion like this begs the question of whether there exists a final theory. In a final theory, we would have laws that we cannot manage to disprove and we would be unable to discover new phenomena that are not described by the theory. The search for such a final theory is the ultimate reductionist goal of physics. With the incredible success of the non-relativistic quantum mechanics of ordinary matter, and the Standard Model of all the subatomic and nuclear forces, it might appear, as many have contemplated, that there could exist a final theory and we might be close to constructing it.

So, is there a final theory and can we construct? I would like to address this question, as we often do in physics, in a geometrical setting. I will describe a geometrical model of knowledge. To construct a model there are at least two requirements. First, the model, should be quantitative and well-defined. The model should explain phenomena that you have already observed or experiments that you have already performed. If the model passes this test then you can have some confidence in it. And then you can use the model to make predictions and suggest new experiments that can be used to test the model. So that's what I'm going to do with knowledge.

The geometrical model that many people have for knowledge is something I call the 'Onion Model'. Whatever it is that you're studying resembles an onion. Scientific research is like peeling back the layers of the onion to get to the core, wherein lies the ultimate understanding. This model, I think, is not very useful and doesn't explain anything. It brings tears to my eyes. Let us turn the onion model inside-out. My geometric model of knowledge resembles an expanding region in a sea of ignorance. We are continuously expanding the realm of knowledge, but always surrounded by ignorance. In this model, we live in a sea of ignorance enclosed in a sphere of knowledge. The sphere of knowledge has been growing very rapidly over the last few centuries, largely due to the application of the scientific method. Most of us are not aware of all of this knowledge, but it is contained in books and journals, and we can all access it now through Google and Wikipedia. At the boundaries of knowledge is ignorance and we pose the questions that push science forward. To paraphrase Donald Rumsfeld, there are the knowable unknowables, namely the ignorance we are aware of, which is limited to the boundary of the sphere of knowledge. But there are also unknowable unknowables, ignorance far from the sphere of knowledge that we are not aware of.

The first thing that is explained by this model is the phenomenon that you are certainly aware of, namely the more you know the more you are aware of your ignorance. Knowledge grows as the volume of the sphere, but ignorance also grows as the surface of the volume of knowledge. This model explains why, even though as time goes on the ignorance that you're aware of grows, that you are wiser — since the volume of knowledge grows faster than the ignorance at its surface. This is a nice model to explain something that is somewhat paradoxical: the more you know, the more you don't know — and yet you are wiser.

Let us use this model to ask whether there a final theory? One way to formulate this question is to ask whether is there a finite amount of ignorance? For if we had a final theory there would be no more ignorance. Could we run out of ignorance? This has happened before, in the exploration of the Earth. There were societies and explorers devoted to exploring and mapping the earth. If the earth had been flat, as some believed, the exploration would have gone on forever, there would always more to explore. But the surface of the earth is not flat, it is the compact surface of a three sphere with a finite surface area. Eventually explorers completed their job and ran out of new territory to explore. Today there is available a rather complete, final map of the Earth, down to distances of tens of meters. Thus, there is a complete theory (or description or map) of the Earth's surface. The analogous question in physics is: Is the sea of ignorance compact?

The sphere of knowledge could be unbounded and infinite, and the amount of ignorance would thus be infinite. We would never run out of ignorance. Knowledge would continue to grow, as would ignorance, and scientific exploration would go on forever. But if the sphere of knowledge is compact, like the surface of the Earth, we will eventually run out of ignorance and complete our final theory. So, what is the answer? I don't know. I'm totally agnostic. It could be either way. I have no idea how to decide. There is, however, an experimental test which might could inform us. In the case of the earth the clue was the curvature of the horizon.

Certainly, at the moment, there is no sign of the curvature that you would see if the knowledge was like the Earth, finite and compact. What would be a sign of curvature? It would be a sign that ignorance is beginning to decrease. As the earth was mapped, at some point, there was very little left to explore. Ignorance of the surface of the earth was decreasing, instead of increasing. That's certainly not the case today, as we seem to have more and more open questions available. Ignorance continues to increase and there is no sign of curvature. But who knows, I am agnostic.

In any case, even if a final reductionist theory does exist its discovery would not be the end of physics, since the structure of the complex collections of simple constituents whose microscopic dynamics are completely understood exhibit phenomena that are new and novel, and exciting and wonderful, and the exploration of their properties can go on forever.

Final theory or not, the questions we pose today appear to be very difficult. Perhaps we are just too dumb to proceed. We are, after all, just educated monkeys that have been around for only a million years or so. We're incredibly arrogant. We've had some successes but we may be too dumb to do certain things, such as survive, or to construct the final theory and answer some of the incredibly difficult questions we're asking today, i.e.: How did the Universe begin? What goes on inside a black hole? We have evolved, and our minds have evolved, to control and understand Nature, but surely not at the level of black holes and quantum cosmology. What survival value does that have? Furthermore, we know that our fellow species are simply too dumb to comprehend even the elementary under-standing that we have developed. I have been totally unable to teach my dog quantum mechanics. I can understand and recognize that a dog cannot grasp quantum mechanics, so why should homo sapiens believe that they are the pinnacle of evolution and are able to grasp, understand and con-struct explanations of the universe in its entirety. We've been doing pretty well so far.

There are some reasons why I think that humans are not limited in their exploration of nature. The distinguishing characteristic of homo sapiens is language, whose most developed form is mathematics, and mathematics has a kind of infinite capacity. In the case of language, Chomsky pointed out that an infant has the innate capability to pick up the rules of language by listening to adults, and then it can formulate sentences that no one has ever formulated before. Thus, there is an infinite capacity of language, es-pecially in its highest form of mathematics. Now, not all infinities are the same, and the infinite capacity of our tools might not be too sufficient to deal with some of the ultimate questions we ask. Again, I don't know. It seems awfully arrogant to say that we can answer anything, but there are experimental signs, or there would be, if we were nearing the point where the deep questions that we ask, such as "How did the universe begin?", were unapproachable by humans. What would be the experimental signs, the danger signs? Well, there probably could be many, but one would be that it would take longer and longer for physics graduate students to get a PhD. Because, if the questions at the boundary of our current knowledge are beyond our capabilities then reaching the boundary would get harder and harder. So, it would take longer and longer to reach the boundary, fi-nally exceeding the lifetime of the graduate student, and that's not the case. As far as I can tell, in the few hundred years we've been asking apparently harder and harder questions, young people, still in their twenties get to the boundaries of our current knowledge and start shaping and answering the questions at the boundaries, just as they have always done. And even if we did discover that we are like dogs, and can't understand the concepts needed to answer some of these questions, we could and most likely will, start dickering with our own brains, or merge with machines and change ourselves, evolve as life always does in order to understand and control the world.

The biggest challenge to the pursuit of fundamental science is that we might lose the will to make the effort and find the means to continue this pursuit. I don't know whether this will transpire. This is a question for those of you who know the laws of history and can predict whether Humanity will continue to explore and devote resources to understand nature. I certainly hope so!

Chapter 4

Are there Ultimate Physical Laws
or are they like the Skins of an Onion

Lars Brink

Department of Physics,
Chalmers University of Technology,
S-412 96 Göteborg, Sweden

Introduction

The modern physics as we know it was started by Johannes Kepler, Galileo
Galilei and Isaac Newton in the 16th and 17th century. We have all read
about the ordeals that Galilei had to suffer since his thesis went against
the Catholic Church. A question arose if there could be laws in Nature that
were independent of a celestial force and that were universal? That became
even more apparent when Isaac Newton formulated his laws for dynamical systems published in his monumental volume Philosophiae Naturalis
Principia Mathematica in 1687. Is the world governed by laws that could
not be manipulated or changed by either human or divine forces? Newton
managed to be a religious man and his contemporary and rival Gottfried
Wilhelm Leibniz stated that "God is an absolutely perfect being". Leibniz
asserted that the truths of theology and philosophy cannot contradict each
other, since reason and faith are both "gifts of God" so that their conflict
would imply God contending against himself.

Already in medieval times thoughts about a **clockwork universe** that
compared the universe to a mechanical clock had appeared. Such a universe continues to tick along, as a perfect machine, with its gears governed
by the laws of physics, making every aspect of the machine predictable.
In a correspondence with Leibnitz the English philosopher and clergyman
Samuel Clarke wrote "The Notion of the World's being a great Machine,

going on without the Interposition of God, as a Clock continues to go without the Assistance of a Clockmaker; is the Notion of Materialism and Fate, and tends, (under pretence of making God a Supra-mundane Intelligence,) to exclude Providence and God's Government in reality out of the World." This was a discussion during the leading intellectuals in Europe that went on over the next centuries and since there is a lot of literature about it I will not go further with this issue. It sets the stage though for the discussion if our world, really our universe, is governed by laws that could not be manipulated or changed and would be true under all circumstances. Another possibility that was not raised much was that the laws were in reality approximate and essentially valid just under certain circumstances. However, nothing seemed to indicate that.

The Perfect World in the Late 19th Century

Isaac Newton's dynamical laws seemed really to work perfectly in every case where they could be used. They not only explained how the apple falls and how the moon can circulate the earth without falling down. It could also be applied to other heavenly bodies. When one studied anomalies in the orbit of the distant planet Uranus one could see that it did not follow Kepler's laws perfectly. Believing in science two astronomers and mathematicians John Couch Adams and Urbain Leverrier computed that there must be yet another planet outside Uranus and in 1846 Neptune was discovered, a great triumph for exact science.

With a lot of patience, time and resources but without computers a number of remarkable calculations were performed to compare with our solar system and the success was always remarkable. Also other branches of physics such as electromagnetism and thermodynamics developed remarkably in the late 19th century with James Clark Maxwell and Ludwig Botltzmann as the great leaders, and the clockwork universe became even more popular. For many a leading scientist it looked obvious that the world followed given laws, and given that it had started one time with given initial conditions the rest could in principle be computed. In the extreme interpretation there could not be a free will. Every human being was a machine that had been started at birth and then followed a scheme that could not be altered. The machine has in reality no free will. All the options we believe we have when we take a decision are just illusions. Our paths are already decided. These ideas were of course not appreciated in the theological world, and were hence mainly ignored.

There were some cracks in the picture though. When computing the orbit of Mercury and taking the interaction with all the other planets into account, it looked as if it did not follow a perfect ellipse. There seemed to be a precession of the perihelium of the orbit of 43 arcseconds per century. Note that one had observed the planet since the time of Galilei, and as I said the scientists had patience and 250 years later the observations were extremely precise. Were there again so far undetected planets or some garbage rotating around the sun that could disturb the orbit of Mercury slightly. The odds for it were overwhelming, but even better and better telescopes could not find anything. Still nobody in the late 19th century doubted the validity of the underlying Newton's equations. It took an Einstein to solve the problem but that was in the next century.

The Einstein Revolution in Physics

When Albert Einstein grew up he reflected whether one could bicycle fast enough to catch up with a light signal. In his annus mirabilis 1905 working as a clerk in the patent office in Bern in Switzerland he solved the problem by first assuming that light travels equally fast in any coordinate system regardless if it is moving in relation to the light source. The velocity of light is a universal constant. Albert Michelson had found this result in experimental studies in the USA some years before, but it is not clear how much Einstein was influenced by this result. With this one assumption he derived 'special relativity' which was an extension of the Newtonian mechanics. The reason why no mechanical experiments or measurements had shown any deviation from Newton's laws was that you can only find such deviations for bodies traveling with velocities relatively close to the velocity of light, $\sim300\,000$ km/s. There had been one indication before that the physical laws were not the final ones. When Hendrik Lorentz investigated the symmetries of Maxwell's equations for electrodynamics he found that they were not the same as the ones underlying Newton's equations. With Einstein's new theory the symmetries were the same and electrodynamics and mechanics were consistent with each other.

Special relativity did not include a description of bodies affected by the gravity force, and it took Einstein another ten years to formulate his General Relativity. This was then the ultimate extension of Newtons laws. In it the gravity force is the effect of space and time being curved. The whole theory is essentially a generalised form of geometry. The first check that Einstein did was to compute the precession of the Mercury orbit and he got exactly the right answer, 43 arcseconds per century. A further check was made in 1919 when the British astronomer Arthur Eddington managed to measure the bending of light from a distant star grazing the sun. This can only be done when there is a full eclipse and in that year one could see one in Africa and in Brazil. Eddington went to Africa and managed to measure the deflection and it was right on the spot. The rest is history. Einstein became a world celebrity touring the world, meeting with the leaders of the world.

So is General Relativity the ultimate law describing the gravity force in the world? So far it has surpassed all hinders to describe the macroscopic universe and it might be the final laws for that. It only differs from Newtonian mechanics when it comes to very strong gravity fields, when large masses move very rapidly and when light passes large masses. We still use Newton's laws to send up satellites or to go to other planets.

General Relativity has also made spectacular predictions like the existence of black holes. These are object which are so dense that the gravity force is strong enough to not let anything out of the object, not even light. The typical example is if we could squeeze the earth to have a diameter of 1 cm. They were for a long time thought to be theoretical dreams but in recent years they have been observed with a clinch this year with the picture of one. This supermassive black hole lies in the centre of the galaxy M87 53 million lightyears away. It hence took the light 53 million years to come here. It weighs 6.5 billion times our sun and has a size of 1 million light years. It is now believed that there are supermassive black holes in the centre of every galaxy. The one in our galaxy, the Milky Way, is 26.000 lightyears away with a mass of 4 million solar masses. There are also strong evidence for black holes around ten times heavier than our sun, remnants of stars that have imploded, and the belief is that they are abundant. This is a very hot subject in physics, and it has become a precision science where General Relativity gives the basic laws that we see the universe to follow.

Is General Relativity hence an ultimate law in Nature? No, we have so far only talked about physics at macroscopic scales and a final law should also describe microcosmos, and most of the developments and new understandings in fundamental physics has occurred there with the quantum revolution.

The Quantum World

The clockwork universe got rocked in its pillars directly at the dawn of the century when Max Planck in 1900 solved the problem with the black body radiation, that is the radiation that an ideal, completely black body radiates. He had to introduce a new constant of nature, $\hbar$, Planck's constant, and assume that the radiation was quantised into an infinite set of oscillators. It took again Einstein's genius to really understand the discovery, when he also in 1905 realised that light is quantised in term of quanta, eventually called photons. (It was for this discovery he finally got the Nobel Prize for 1921.) In the mid-1920's Werner Heisenberg and Erwin Schrödinger then revolutionised fundamental physics by introducing quantum mechanics. This was a completely new framework very different from the old Newtonian mechanics of the macroworld. Every event that happens can only be described by statistics. Consider a sample of radioactive uranium atoms. We cannot set up a scheme beforehand when there will be a decay. We can only say how many on average we will have per hour. So the microworld is not a clock! Quantum mechanics goes over to the old classical mechanics for macroscopic systems, and we recover the macroscopic laws I have talked about above. They are clearly not the fundamental laws then!

For the last 100 years the basic physics research has dug deeper and deeper into the quantum world. The first fundamental force to be understood was the electromagnetic one with the formulation of Quantum Electrodynamics in the 1940's. Again this theory goes over to Maxwell's laws for macroscopic systems. This theory underlies condensed matter physics and atomic physics and has revolutionised the modern world, where most gadgets we use today use quantum properties understood from basic laws. In the last fifty years we have also got an understanding of the nuclear forces, the strong one keeping the nuclei together and the weak one leading to radiative decays. The three forces are united into the Standard Model of Particle Physics which works perfectly at the scales we can probe now with the big accelerators, where we can check physics down to 10^{-18} cm.

How about quantum mechanics and the gravity force? Einstein's theory cannot be taken over straight to the quantum world like Maxwell's electrodynamics. The most promising attempt is the Superstring Theory, where the fundamental constituents are string-like objects instead of point-like as in the quantum mechanics of today. Superstring Theory includes quantum gravity with the other fundamental forces but so far we have not been able to find a unique such theory. The theory is based on deep and possibly new mathematics, and we have so far only scratched the surface. This could be the ultimate theory and in this sense there could be an absolute truth. Only the future research will show if we are on the right track.

Physical Laws as the Skin of an Onion

In his beautiful and deep essay "Nature Conformable to Herself: Arguments for a Unified Theory of the Universe" the late Murray Gell-Mann, Nobel laureate in physics in 1969, described progress in fundamental physics as a progression of understanding through layers of phenomena characterised by higher and higher energy scales and less and less experimental accessibility. He used the metaphor peeling away the skin of the onion to characterise this progression.

Einstein once said that the most incomprehensible thing about the world is that it is comprehensible. The most outer skin of our comprehension is represented by the laws or regularities that we see around us in the world, how large numbers of people behave, how our landscape changes and on the largest scales how our universe is behaving. When we peel off that skin and perhaps a few more we come to the skin where Newton's and Maxwell's laws are the correct ones to describe the phenomena at that scale. We peel further and go down in length scales to where Quantum Electrodynamics is the correct law explaining the physics that we see at those scales. We also there could set up laws how the protons and the neutrons of the atomic nuclei behave, how they make up the nuclei and also show the variety of nuclei. At the next layer we learn of new laws telling us that the protons and the neutrons are in fact built up by the quarks invented by Gell-Mann and George Zweig, and we are led to the Standard Model mentioned above. There might be further layers before we reach the Superstring Theory if that is the correct theory to join in gravity at very small length scales. Can this be the last skin, the centre of the onion? Perhaps it is presumptuous by me to believe that it is a logical possibility. Perhaps there are further layers never ending . . . . Only future research will tell us.

Suppose now that there are ultimate laws describing the world. What can they tell us about our macroworld? Very little. If we put back the skins, they do not talk much to each other. We have to rediscover the laws at every level. None of these are then absolute, but adequate and general laws at those scales. The further out we go the laws become more and more approximate and are obviously not laws in a strict meaning but more valuable relations.

Even if we loose rigour when going to an outer skin we might imagine that with enormous computing power we could derive the laws of the outer skin from the ones inside. We do not know that. There could also be unsurmountable obstacles that will prevent us from doing so.

I have argued that a final theory could be found which should constitute an absolute truth in a certain sense. Does it then lead to a clockwork universe? We have to remember that the results of the theory are statistical, and that might leave room for a free will. Whatever future scientists will find I think we should go on to continue to decide about our lives.

Chapter 5

The CERN Model, a Collective Machinery
to Test Laws Against Nature

Michel Spiro* and Maurizio Bona

CERN

With the generic term "CERN model" it is intended for the ensemble of official documents, organizational structures, managerial culture, and behaviour of people including the user community, which allows the successful operation of such a particular intergovernmental Organization and research institution as CERN. This paper will address the main elements on which the CERN model, a collective machinery to test laws against Nature, is based, and propose some ideas why the CERN model might inspire and possibly be adopted in other fields.

1. The CERN Convention

An important element of the "CERN model" is the Convention, signed in 1953, which established CERN as an intergovernmental Organization by treaty. The Convention, in addition to the necessary legal provisions for such an official text, provides a strong definition of the main guideline the organization is to follow, which is summarized in the following sentence: *"The Organization shall have no concern with work for military requirements and the results of its experimental and theoretical work shall be published or otherwise made generally available"*. Indeed CERN, throughout its more than 60 years of existence, made the transparency of its activities, the sharing its results, and the encouragement and implementation of open collaboration, the main signs of its culture.

---

* Lecturer

Moreover, the CERN Convention, without missing its role of fundamental document for an intergovernmental Organization, provides an effective set of provisions that are necessary for the operation of a scientific research centre, whose main objective is not to discuss science but actually to do it successfully.

The value of the CERN Convention is further confirmed by the fact that it inspired the Conventions of other scientific intergovernmental Organizations, e.g. the conventions of the European Space Agency (ESA), the European Southern Observatory (ESO) and, more recently, of SESAME, the Synchrotron-Light for Experimental Science and Application in the Middle East, the Joint Institute for Nuclear Research in Dubna (former USSR member states).

2. Politics and Research

CERN was established on the ashes of World War II by a group of inspired scientists and diplomats who understood the importance to promote cooperation among former antagonists and to contribute to post-conflict integration through science. The success of this joint initiative between scientists and diplomats and the convinced support given by the CERN Member States throughout the years, show that politics and science have been able to find converging interests and to exploit synergies.

However — and this is a key element of the "CERN model" — although present in the CERN Council via representatives of the Member States, politics has been capable, thanks to the mutual trust built up with the world of science, to avoid interfering with the strategic scientific choices and programmes of the Organization.

A significant example of this attitude is the Scientific Policy Committee, one of the two main Committees of CERN's Council (the other one is the Finance Committee). Members of the Scientific Policy Committee are not elected by the Member States, but are co-opted by the Committee members themselves. Moreover, some of them are not even citizen of any of CERN's Member States. Therefore, the outcomes of the work of the Committee and its proposal to the Council represent the expression of the real scientific interests of the international particle physics community, not influenced by any political concern.

3. CERN as a Platform for the International Scientific Community

Another characteristic of the CERN's culture is its open approach to the whole scientific community involved in particle physics research, including to non-Member States. Besides about 3000 staff members on its payroll, CERN counts at present more than 12,000 researchers, whose salaries are paid by external research institutions, coming from all over the world, including from non-Member States.

This means that the research infrastructure developed and maintained by CERN (e.g. its accelerators) are made available to the entire inter-national scientific community. This worldwide community participates in the experiments foreseen by CERN's scientific programme, through Collaborations in which CERN contributes only as one of the involved research institutions, both in terms of personnel and budget, the rest being covered by national research institutions and Universities, but nevertheless acts as a conductor of an orchestra.

CERN makes a lot of effort in cooperating with and supporting the national research communities that operate on particle physics. This ensures the effective coordination between the global (international) and the local (national) levels of action, which is indispensable for advanced countries to provide their highest possible contribution, and for less advanced countries (in terms of scientific know-how) to reduce the gap with the other countries. Therefore, openness and inclusion, associated to the promotion of the combination between the global and the local levels, are essential elements in CERN's approach to the implementation of its scientific programmes.

4. The Scientific Collaboration Culture

Experimental collaborations at CERN bring together very large numbers of researchers from all over the world. Two out of the four big experimental Collaborations that operate detectors at CERN's Large

Hadron Collider (LHC) count up to 3000 researchers coming from more than 175 research institutions and universities from dozens of countries. All these people converge to CERN with their cultures, ambitions and, sometimes, big egos.

However, once at CERN they have to leave all possible personal objectives apart, and rather concentrate on the best way to achieve — all together — the common scientific objective for which the Collaboration was set up. This sharing of a superior objective that reduces any personal objective to irrelevance, the absolute need to interact openly and trustfully with the other members of the Collaboration, allows achieving important scientific results without heavy hierarchical structure in the Collaborations, with decisions based as widely as possible on consensus, although debate and exchange of opinions is not only tolerated but encouraged. In such a context, cultural differences become assets and not obstacles as is sometimes the case in other fields where political considerations predominate.

5. Management and Hierarchy

The CERN budget is practically 100% regular budget, with contributions from the Member States that are proportional to their GDPs. Other contributions (EU, sponsors, special contributions ...) do not exceed 10% of the total budget. All the money is put in the same pot and managed by the CERN management, under the monitoring of the Council, thus avoiding the dispersion that is sometimes present in other organisations where a consistent part of the budget comes from voluntary contributions.

Careers at CERN are based on merit and, in particular at the mid-high levels, on recognition by peers. This does not require a heavy managerial structure, and hence results in a situation that favours flexibility and rapid decision-making processes. A relatively light managerial and decisional structure allows quick and effective decisions, and also permits the younger staff or those in the lower grade to concretely influence the work of the Organization, provided of course their arguments and proposals are solid.

However, there is a difference between the management of the 3000 CERN staff members and the 3000 scientists distributed collaborations. In the case of CERN, the responsible persons are designated by the Director General (under the supervision of Council) while they are elected in the case of large experimental collaborations. But like in the case of the experimental Collaborations, also CERN staff members (i.e. the individuals on CERN's payroll) usually show a high level of sharing of the overall objectives of the Organization and of the department/group they work for. They all signed up (collaboration members and CERN staff) a shared code of conduct. This eases the management of the personnel, and also contributes to creating the culture of openness and the natural tendency to implement trustful cooperation without boundaries, which constitutes the CERN culture. Therefore CERN, in both its components as an intergovernmental Organization and as a research centre, is naturally predisposed to promote international peaceful cooperation and to establish bridges among cultures.

6. Science, Technology, Innovation and Education

CERN mandate is to carry out scientific programmes on particle physics. In order to accomplish its mandate successfully, CERN must master sophisticated technology, which requires continuous innovation. Therefore, the Organization has to carefully conjugate science with technology and innovation, investing adequately in all the different components in order to maintain and further develop a virtuous circle. CERN is open to new ideas and to contributions on possible new developments coming from inside the Organization as well as from the various external partners. The spirit of openness and trust mentioned above creates a fertile ground to develop new ideas and make them concrete.

CERN, thanks to the support of its member states and the support of the global scientific community succeeded to establish with unprecedented precision the Standard Model of particle physics, to discover the missing pieces, the mediators of the weak interaction and to the discover the Higgs boson at the origin of the mechanism which gives mass to elementary particles. These achievements lead to a new level of knowledge, the physics that govern matter at scales of 10^{-16} cm down to 10^{-18} cm.

Meanwhile, many innovations that changed the world were stimulated by CERN: first touch screens, birth of the world wide web at CERN, cloud computing, PET tomography, high field magnets for MRI, use of accelerators to treat tumours, open source software and now open access publishing and open innovation.

CERN illustrates very well the EU target: open science, open innovation, open to the world.

Continuous development of accelerator and detector technologies is indispensable to tackle the challenges of its scientific programmes. Studies and developments on new technologies to be adopted in the next generation's accelerator or particle detectors often begin as soon as the accelerator or particle detectors of the present generation start their operation. Planning the future in the particle physics world is extremely difficult, as the tendency of the major international projects is to become bigger and bigger and take more and more time to be developed and completed (about 25 years for the CERN Large Hadron Collider, the LHC). Therefore, an essential element for CERN when it prepares a new project is to correctly identify the objectives, the technological challenges, the human and financial needs, and the road map of the different required steps.

In this respect, the precise subdivision of the work between the different groups/institutions involved, the definition of the required coordination structure(s) and the relevant interfaces, the capability to maintain clear the visibility of the overall objectives, are key elements of CERN's know-how on the design, preparation, and execution of big challenging projects. For this reason, CERN is often proposed as an example to follow to set up new international scientific research infrastructures (e.g. OECD Report[†]).

Also, education is an integral part of the CERN model. Science, Technology, Engineering and Mathematics education, also referred to as

---

[†] "Report on the Impacts of Large Research Infrastructures on Economic Innovation and on Society — Case Studies at CERN", OECD, September 2014.

STEM education, is a vital element to maintain the virtuous cycle of CERN. Without being a University or a traditional education centre, CERN hosts a large number of students in physics, engineering, IT and other disciplines, from undergraduate to PhD level, for periods varying from a few months to a few years. The main characteristic of CERN's educational programmes is that the students not only receive theoretical training on the most recent developments in the different fields concerned, but are integrated into the real everyday work of groups and units that carry out the programmes of the Organization. CERN also provides training to teachers, in particular to high school teachers, on the most recent developments in physics and on some effective didactical models developed at CERN. The teachers' training programme, which started for teachers from CERN Member States, is now offered also to teachers from non-Member States, reaching a total of more than 1000 teachers trained every year.

7. European Strategy for Particle Physics

Every five or six years, CERN orchestrates the update of the European Particle Physics strategy. It is a two-year bottom-up process ratified at the end by the CERN Council, taking into account political and economical constraints. Again, like for the rest, bottom-up and top-down approaches are tentatively and in the end successfully reconciled.

The strategy develops an updated long-term vision (the future of CERN for the next 50 years or more is being discussed now) together with short-term implementation and middle-term preparation.

It implies a global effort distributed locally in work packages, CERN playing the role of a conductor and a major actor.

8. Conclusive Remarks

The "CERN model" is a mix of documents, managerial solutions and cultural approaches that developed over the 60 years of history of the Organization.

Its success resides in the simultaneous presence of a number of elements, the main of which can be summarized as follows:

- A Convention that defines clearly the objectives and the mandate of the Organization, and which is open to cooperation with non-Member States and other external partners.
- A relatively light managerial structure, but which is fully responsible to the Council of the resources invested by the Member States, ensures effective decisional processes combined with adequate adaptation capabilities.
- The mutual respect and the collaborative attitude existing between the scientific component of CERN, which is the expression of the international scientific particle physics community, self-managed with a rather flat hierarchy, and the political component that represent the interests of the Member States.
- The good coordination and the synergies existing between the activities and programmes carried out at CERN (the "global" level), and those carried out at the national levels by the national research institutions and Universities.
- An open-minded culture, based on shared values like merit, openness, trust, peaceful cooperation, respect and inclusion, and the subsequent code of conduct.
- The coexistence in the same groups/units of people of all ages and experiences, ranging from the undergraduate student to the famous professor or the Nobel prize laureate.
- Coopetition: there is a mixture of collaboration and competition (emulation) within the community that makes CERN rather unique. How can 3000 scientists collaborate in a single experiment and still compete with each other to be recognized as an eminent scientist? This can only be achieved because they all share the main goals and because the individual success is based on peer merit recognition.
- The capability to maintain and foster a virtuous circle involving science, technology, innovation, and education.
- The capability to identify the overall objectives that all the people working at CERN must share, and to subdivide complex projects into sub-parts, with effective coordination and interface structures.

- In summary, the CERN organization at large reconciles bottom-up and top-down approaches, long-term vision and short-term implementation, global effort distributed in local work packages, collaboration, and emulation. It is a machinery to increase our global knowledge and public good. It is worth considering how it could be applied to other common global public goods.

© 2025 World Scientific Publishing Company
https://doi.org/10.1142/9789811292316_0006

Chapter 6

The Argument Against Quantum Computers, the Quantum Laws of Nature, and Google's Supremacy Claims

Gil Kalai*

The Hebrew University of Jerusalem and IDC, Herzliya, Israel

My 2018 lecture at the ICA workshop in Singapore dealt with quantum computation as a meeting point of the laws of computation and the laws of quantum mechanics. We described a computational complexity argument against the feasibility of quantum computers: we identified a very low-level complexity class of probability distributions described by noisy intermediate-scale quantum computers, and explained why it would allow neither good-quality quantum error-correction nor a demonstration of "quantum supremacy," (a.k.a. "quantum advantage"), namely, the ability of quantum computers to make computations that are impossible or extremely hard for classical computers. We went on to describe general predictions arising from the argument and proposed general laws that manifest the failure of quantum computers.

In October 2019, *Nature* published a paper[1] describing an experimental work that took place at Google. The paper claims to demonstrate quantum (computational) supremacy on a 53-qubit quantum computer, thus clearly challenging my theory. In this paper, I will explain and discuss my work in the perspective of Google's claims.

1. Introduction

In this paper I want to present to you my theory explaining why computationally superior quantum computing is not possible, discuss the laws of nature that may support this theory, and describe some potential connections and applications. This is a fairly ambitious task; for one, many experts do not understand my argument, and even more do not agree with me. On top of that, the assertion of a paper[1] published in *Nature* in October 2019, declaring that "quantum computational supremacy" was achieved by a team from Google on a 53-qubit computer, seems to refute my argument.

*Work supported by ERC advanced grant 834735.

We will describe and give a preliminary evaluation of Google's claims. The story of quantum computers is related to exciting developments and problems in physics and in the theory of computation, and my purpose here is to tell you about it in non-technical terms (with short subsections entitled "under the mathematical lens" that offer a glimpse of the mathematics and can be skipped).

1.1. *Paper outline*

Sections 2 and 3 introduce classical and quantum computation. Among other things, we discuss an important heuristic concept of "naturalness" that is at the heart of the interface between computational complexity and the practice of computing. Most of the paper is devoted to three related topics. The first is my argument laid out in Sec. 4 of why quantum error-correction and quantum advantage are not possible. The second is a description of general laws of nature that emerge from the failure of quantum computers and quantum error-correction. Those are described in Sec. 5 and further connections are given in Sec. 8. The third is a study of the Google supremacy claims. Following a description of these claims in Sec. 6, we adduce in Sec. 7 reasons for thinking that the Google claims are not reliable and discuss how to further study them.

In Sec. 9 we briefly discuss developments that occurred since the first version of the paper was written. In particular, a 2020 paper[2] in *Science* claimed an even greater quantum computational advantage using a photonic system operating at room temperature. The authors of Ref. 2 claim that their device provides in 200 seconds samples that would require a classical computer billions of years. However, it turns out that the statistical argument from Ref. 2 is incorrect, and this is closely related to my general argument against quantum computers. My 2014 paper with Guy Kindler[3] showed a simple model demonstrating how a classically sampled distribution may pass the same statistical tests by only reproducing small-scale correlations of the actual theoretical distribution.

A quick remark about terminology. The terms "quantum supremacy" and "quantum advantage" are synonyms and are both used to refer to the ability of quantum computers to make computations that are impossible or extremely hard for classical computers. When we refer to works by others we will adopt the terminology used in those works but, in view of a recent critique of the term "quantum supremacy," we will otherwise use the term "quantum advantage." ("Quantum advantage" always refers to a computational advantage by several orders of magnitude, and I proposed the term "huge quantum computer advantage," or "HQCA" for short that captures the expected magnitude of quantum advantage for some computational tasks.) In any case, a main message of this paper is that quantum supremacy, however referred to, is not possible.

Acknowledgment: I am thankful to Yosi Atia, Ramy Brustein, Guy Kindler, Eliezer Rabinovici, Jelmer Renema, Yosi Rinott, Tomer Shoham, and Barbara Terhal for helpful discussion.

2. Classical Computers

2.1. *Easy and hard problems*

The central concept in the theory of computational complexity is that of an *efficient algorithm* (also called "polynomial-time algorithm"). An efficient algorithm is an algorithm that requires a number of operations that is at most *polynomial* in the size of the input. The class of algorithmic tasks that admit efficient algorithms is denoted by **P**. For example, given a list of n numbers, the task of finding the maximal number has an efficient algorithm.

Another important algorithmic task is that of matching. Let me elaborate a little: we are given two collections A and B of an equal size n, and for every element $a \in A$ we are given a set $B_a \subset B$. The task at hand is to decide whether we can find a function f from A to B such that

- $f(a) \neq f(a')$ for every distinct a and a',
- $f(a) \in B_a$ for every a.

Such a function is called a *perfect matching*. A major landmark in computer science was the discovery by Ford and Fulkerson of an efficient algorithm for matching.

Our third algorithmic task will be the famous *traveling salesman problem*. There are n cities and on the road between each pair of cities c_1 and c_2 there is a toll $T(c_1, c_2)$. A traveling salesman needs to travel between these cities, that is, to start at city c_1 and then to travel through each city exactly once, until returning to the initial city, so as to minimize the overall toll. There is a simpler version of this problem that is called the *Hamiltonian cycle problem*. For every pair of cities c_1 and c_2 we are told in advance whether the road connecting the two is open or closed. The challenge is to start at city c_1 and then to travel through each city exactly once, returning at the end to c_1 and using only open roads. Such a route is called a Hamiltonian cycle. A major conjecture in the theory of computational complexity is that there is no efficient (polynomial-time) algorithm for solving the traveling salesman problem and there is no efficient algorithm to tell whether a Hamiltonian cycle exists. In fact, it is commonly believed that an algorithm for these problems (in the most general cases) requires an exponential number of steps and therefore goes beyond the reach of digital computers, as the number of cities grows. The task of deciding whether there exists a Hamiltonian cycle constitutes an **NP-complete** problem: being in the computational class **NP** means that there is proof that a graph G has a Hamiltonian cycle that can be verified in a polynomial number of steps. Being **NP-complete** means that any other **NP**-problem can be reduced to this problem.

2.2. *When theory meets practice: Naturalness in computer science*

Our main tools for the study of the complexity of algorithms are asymptotic. For example, we make a distinction between exponential running time and polynomial running time. When trying to gain insights into practical questions we need to make an assumption of *naturalness*, namely, that the constants involved in the asymptotic descriptions are mild. Without such an assumption, computational complexity insights hardly ever apply to real-life situations. With the assumption of naturalness we do gain much insight: if an algorithmic task can be solved (asymptotically) in a polynomial number of steps, then usually this suggests that the task is practically feasible. On the other hand, if a class of algorithms, or computational devices, represents polynomial-time computation, then usually we cannot expect that this class of algorithms will practically solve intractable problems. For example, if we are offered a device for solving the Hamiltonian cycle problem, and we can analyze the device and realize that it represents an asymptotically polynomial-time algorithm, then we cannot expect that this device will outperform, by a very large margin, ordinary digital computers. Naturalness is a heuristic assertion, but it is a powerful one. Of course, the lower the computational power of a class of algorithms or computing devices is in the hierarchy of computational complexity classes, the more implausible it becomes that such algorithms or computing devices will allow, in practice, powerful computation.

2.3. *Randomness and computation*

One of the most important developments in the theory of computing was the realization that the addition of an internal randomness mechanism can enhance the performance of algorithms. Since the mid-1970s, randomized algorithms have become a central paradigm in computer science. One of the greatest achievements was the polynomial-time randomized algorithms of Solovay and Strassen (1977) and Rabin (1980) for testing whether an n-digit integer is a prime. Rabin's paper stressed that the algorithm was not only theoretically efficient but also practically excellent, and gave "probabilistic proofs" that certain large numbers, like $2^{300} - 153$, are primes. This was a new kind of proof in mathematics.

2.4. *Under the mathematical lens: Determinants and Lovasz's algorithm for perfect matching*

Let us go back to the problem of finding a perfect matching and consider an n-by-n matrix M where the rows correspond to the elements of A, $a_1, a_2, \ldots, a_n$ and the columns correspond to the elements of B, $b_1, b_2, \ldots, b_n$. Now we consider variables x_{ij} for every $i, j, 1 \leq i \leq n$, $1 \leq j \leq n$, and let $m_{ij} = 0$ if $b_j \notin B_{a_i}$ and $m_{ij} = x_{ij}$ if $b_j \in B_{a_i}$. Lovasz's first observation was that a perfect matching exists if and only if the determinant of M (regarded as a polynomial in the variables x_{ij}s) is not zero. Lovasz's second observation was that if you create a new matrix M' by replacing x_{ij} with a random element in a large finite field, and if the determinant of M is not zero, then, with high probability, the determinant of M' is not zero either.

Here is Lovasz's algorithm: given the data, we build at random the matrix M' and check whether its determinant equals zero and repeat this process k times. If we get a non-zero answer once, we know that there is a perfect matching; if we always get zero, we know with high probability that a perfect matching does not exist.

We need one additional ingredient that goes back to Gauss: when the entries are concrete numbers, there is a polynomial-time algorithm for computing determinants. This is based on Gauss's elimination method, and can be considered as one of the miracles of our world.

3. Quantum Computers

3.1. *Huge computational advantage: Factoring and sampling*

Quantum computers are hypothetical physical devices that allow the performance of certain computations well beyond the ability of classical computers, in a polynomial number of steps in the input size. The basic memory unit of a quantum computer is called a *qubit* and the basic computational step on one or two such qubits is performed by *gates* (further details are given below). Shor's famous algorithm[47] shows that quantum computers can factor n-digit integers efficiently, in roughly n^2 steps! (The best known classical algorithms are exponential in $n^{1/3}$.) This ability for efficient factoring allows quantum computers to break the majority of current cryptosystems.

A *sampling task* is one where the computer (either quantum or classical) produces samples from a certain probability distribution D. In the main examples of this paper each sample is a 0-1 vector of length n, where D is a probability distribution on such vectors. Quantum algorithms allow sampling from probability distributions well beyond the capabilities of classical computers (with random bits). Shor's algorithm exploits the ability to sample efficiently on a quantum computer a probability distribution based on the Fourier coefficients of a function.

3.2. *Noisy quantum computing*

Quantum systems are inherently noisy: we cannot accurately control them, nor can we accurately describe them. In fact, every interaction of a quantum system with the outside world amounts to noise. A noisy quantum computer has the property that every computational step (applying a gate, measuring a qubit) makes an error with a certain small probability t. (These errors are described more specifically in Sec. 6, whereas in Sec. 5.5 we get a glimpse of the mathematics of noise in quantum systems.) The threshold theorem[4-6] asserts that if the rate of errors t is small enough (and if a few additional assumptions are made), then a noisy quantum circuit can simulate noiseless quantum circuits. To implement such a simulation we need certain building blocks called *quantum error-correcting codes,*[48,49] where a collection of 100–5000 quantum qubits (or more) can be "programmed" to represent a single stable "logical" qubit.

3.3. *NISQ computers*

Noisy intermediate-scale quantum (NISQ) computers, are quantum computers where the number of qubits is in the tens or at most in the hundreds. Over the past decade researchers have conjectured[7] that the huge computational advantage of sampling with quantum computers can be realized by NISQ computers that only approximate the target probability distribution. These researchers have predicted that quantum computational advantage (for sampling tasks) could be achieved for NISQ computers without using quantum error-correction. NISQ computers are also crucial to the task of creating good-quality quantum error-correcting codes. An important feature of NISQ systems — especially for the tasks of achieving quantum advantage and quantum error-correction — is the fact that a single error in the computation sequence has a devastating effect on the outcome. In the NISQ regime, the engineering task is to keep the computation error-free. We shall refer to the probability that not even a single error occurs as the *fidelity.*

Many companies and research groups worldwide are implementing quantum computations via NISQ computers (as well as by other means). There are several different approaches to realizing individual qubits and gates, and each of the main approaches is marked by different variations. Realizing quantum circuits by superconducting qubits is a leading approach, whereas trapped-ion qubits, photonic qubits, topological qubits, and others are considered notable alternatives.

3.4. *Under the mathematical lens: The mathematical model of quantum computers*

3.4.1. *Quantum computers (circuits)*

- A *qubit* is a piece of quantum memory. The state of a qubit is a unit vector in a two-dimensional vector space over the complex numbers $H = \mathbb{C}^2$. The memory of a quantum computer (quantum circuit) consists of n qubits and the state of the computer is a unit vector in the 2^n-dimensional Hilbert space, i.e., $(\mathbb{C}^2)^{\otimes n}$.
- A *quantum gate* is a unitary transformation. We can put one or two qubits through gates, which represent unitary transformations, that act on the corresponding two- or four-dimensional Hilbert spaces. There is a small list of gates that are sufficient for the full power of quantum computing.
- *Measurement* of the state of k qubits leads to a probability distribution on 0-1 vectors of length k.
- A *quantum circuit* is composed of a collection of gates acting successively on n qubits. To describe an efficient (or polynomial-time) quantum algorithm, we assume that the number of gates is at most polynomial in n. (We also assume that the sequence of gates can be produced efficiently by a classical algorithm.)

3.4.2. *Superposition and entanglement*

The state of a single qubit is a *superposition* of basis vectors of the form $a|0\rangle + b|1\rangle$, where a, b are complex and $|a|^2 + |b|^2 = 1$. The complex coefficients a and b are called *amplitudes*. A measurement of a qubit in state $a|0\rangle + b|1\rangle$ will lead to a random bit of 0 with probability $|a|^2$ and 1 with probability $|b|^2$. This rule for moving from complex amplitudes to probabilities is referred to as the "Born rule."

Two qubits are represented by a tensor product $H \otimes H$ and we denote $|00\rangle = |0\rangle \otimes |0\rangle$. The *cat state* $\frac{1}{\sqrt{2}}|00\rangle + \frac{1}{\sqrt{2}}|11\rangle$ can be regarded as a quantum analog, called *entanglement*, of correlated coin tosses that yield two heads with probability 1/2, and two tails with probability 1/2. The cat state is the simplest example of entanglement, and the strongest form of entanglement between two qubits.

4. The Argument Against Quantum Computers

4.1. *My argument against quantum advantage and quantum error-correction*

Here, in brief, is my argument against quantum computers. For more details see Refs. 8 and 9.

> **(A)** From the perspective of computational complexity theory, noisy intermediate-scale quantum (NISQ) circuits are low-level classical computational devices.

> **(B)** Therefore, by naturalness, NISQ systems do not support quantum advantage. In other words, the rate of noise cannot be reduced to the level allowing quantum advantage.

> **(C)** Achieving good-quality quantum error-correction requires an even lower noise rate than the one required for achieving quantum advantage.

> **(D)** Therefore, NISQ systems do not support quantum error-correction.

> **(E)** Hence, large-scale quantum computing based on quantum error-correction is beyond reach.

4.2. *Four thresholds*

To put the above argument a little differently, we can consider four crucial thresholds of noise, $\alpha, \beta, \gamma, \delta$:[a]

- α is the rate of noise required for universal quantum computing,
- β is the rate of noise required for good-quality quantum error-correction,
- γ is the rate of noise required for quantum advantage, and
- δ is the rate of noise that can realistically be achieved.

Since universal quantum computing requires very good-quality quantum error-correcting codes, we get that $\alpha < \beta$. At the center of my analysis

---

[a]α, β, γ, and δ are not universal constants; they depend (moderately) on a specific implementation. Our argument asserts that inequality (1) holds universally.

is a computational complexity argument stating that $\gamma < \delta$, and I also rely on the argument that $\beta < \gamma$, which is in wide agreement. Given these inequalities, we get that

$$\alpha < \beta < \gamma < \delta. \tag{1}$$

We note that it is a strong intuition of many researchers that with sufficient engineering efforts, δ can be reduced as close to zero as we want. My argument implies that this belief is incorrect.

4.3. *Four facts that strengthen the argument*

There are four facts that strengthen this argument against quantum computers.

1. The first is that NISQ circuits are very, very low-level classical computational devices.
2. The second is that while our argument asserts that the level of noise that can realistically be achieved will be above the level of noise allowing the demonstration of quantum advantage, there is yet another, related argument asserting that when we consider n-qubit circuits, then for a wide range of lower levels of noise, the outcomes will be *chaotic*: no robust probability distributions will be possible as the output.
3. The third fact is that there are also direct reasons why probability distributions supported by quantum error-correcting codes (like the popular "surface codes") are not supported by the very low-level computational complexity class of NISQ circuits.
4. The fourth fact is that while quantum error-correction requires achieving very high fidelity for tens or hundreds of qubits, it has been realized in recent years (and this forms the very basis for Google's experiment) that quantum advantage can be demonstrated even with low fidelity.

The first and second items in the list are the most important, and I would therefore like to say a little more about them. (The reader is referred to our next mathematical Sec. 4.4 and to Refs. 8 and 9 for more details.)

The computational complexity class describing NISQ circuits is **LDP** (low-degree polynomial) and this class is contained in the familiar class of distributions that can be approximated by bounded-depth (classical) computation.

Let me phrase the second point a little differently. The threshold for realistic noise δ cannot be pushed down to allow quantum advantage; but more than that is true: there is a large range of error rates below δ, where even if you could reduce the error rate to these levels, the resulting probability distribution would be chaotic and would largely depend on the fine parameters of the noise itself.

4.4. *Under the mathematical lens: Noise stability and sensitivity and Fourier–Walsh expansion*

The first assertion in my argument is related to a mathematical theory of noise stability and noise sensitivity that goes back to Benjamini, Kalai, and Schramm (1999)[10] (and can be traced back to Ref. 11). In my lecture in Singapore I described this theory in the context of voting methods. How likely is it that the outcome of an election will be reversed because of noise in counting the votes?

Let Ω_n be the set of 0-1 vectors of length n. We start with a real function $f(x_1, x_2, \ldots, x_n)$, and for a real number t, we define the noise version of f as

$$N_t(f)(x) = \sum_{y \in \Omega_n} f(x+y) t^{|y|} (1-t)^{n-|y|}. \tag{2}$$

Here $y = (y_1, y_2, \ldots, y_n)$ is also a 0-1 vector and $y_i = 1$ indicates "error in the ith coordinate." The sum $x+y$ should be considered as a sum modulo 2: $x_i + 0 = x_i$ and $x_i + 1 = 1 - x_i$, and $|y| = x_1 + x_2 + \cdots + x_n$.

It turns out[10] that the behavior of noise for functions on Ω_n is closely related to the Fourier–Walsh expansion of the function. Here is a quick description. Recall that for $S \subset [n] = \{1, 2, \ldots, n\}$, the Walsh function W_S is defined as

$$W_\emptyset = 1 \quad \text{and} \quad W_S(x_1, x_2, \ldots, x_n) = \prod_{i \in S} (1 - 2x_i). \tag{3}$$

If the Fourier–Walsh expansion of f is

$$f = \sum_{S \subset [n]} \hat{f}(S) W_S, \tag{4}$$

then

$$N_t(f) = \sum_{S \subset [n]} \hat{f}(S)(1 - 2t)^{|S|} W_S. \tag{5}$$

If the value of f is always 0 and 1 we call f a Boolean function, then we can regard f as a voting rule for a two-candidate election. A deep finding from Ref. 10 is that for a wide class of voting rules, only those voting rules that are close enough to the "majority" voting rule (or a weighted version of the majority rule) are noise-stable. We note that the majority voting rule is related to the very basic methods for achieving robust classical information and computation.

We can now describe the "very low-level" computational complexity class **LDP** of probability distributions described by NISQ systems. The class **LDP** consists of probability distributions that can be approximated by polynomials of bounded degree. Indeed, when $t > 0$ is fixed and we apply the noise N_t (defined by Eq. (2)) to an arbitrary probability distribution D, the resulting distribution $N_t(D)$ can be well approximated by polynomials of bounded degree (roughly $1/t$). (This easily follows from Eq. (5).) Distributions that can be (approximately) described by bounded-degree polynomials can also be approximately described by *bounded-depth (classical) circuits*. (Bounded depth circuits define a well-known low-level complexity class $\mathbf{AC^0}$.) Approximate sampling for **LDP**-distributions is also *efficiently learnable*, which also describes a low level complexity class for approximate sampling. (See Fig. 1).

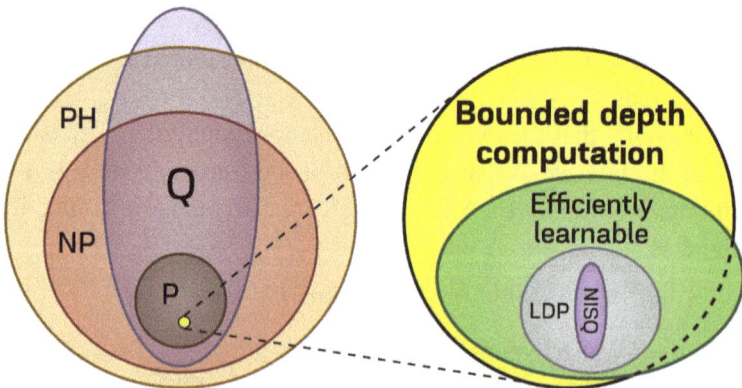

Fig. 1. Probability distributions described by NISQ systems represent a low-level computational class **LDP**. Approximate sampling for **LDP**-distributions belongs to the class of bounded depth (classical) computation and is also efficiently learnable.

When D is a probability distribution proposed for "quantum advantage" (or arising from quantum error-correcting codes), then, even when the level of noise is subconstant but (well) above $1/n$, the correlation between the two distributions D and $N_t(D)$ tends to zero. This suggests that for realistic forms of noise the noisy probability distribution will strongly depend on fine parameters of the noise itself, leading to a "chaotic" behavior.

We note that the analysis of noise sensitivity of NISQ systems was initially carried out on another model called "boson sampling" in Ref. 3. For further discussion of boson sampling, see Refs. 7–9,12,13, and Sec. 9.

5. The Laws

Without further ado let us now move to the laws of physics that emerge from the failure of quantum computers.

Law 1: Time-dependent quantum evolutions are inherently noisy.

Law 2: Probability distributions described by low-entropy states are noise-stable and can be expressed by low-degree polynomials.

Law 3: Entanglement is accompanied by correlated errors.

Law 4: Quantum noise accumulates.

We emphasize that these four laws are compatible with quantum mechanics. The laws proposed in this section are not part of the argument for why quantum error-correction is not possible, but largely rely on taking this argument for granted.

5.1. *The first law: Time-dependent quantum evolutions are inherently noisy*

Time dependence in a quantum evolution amounts to an interaction with the environment and the first law asserts that there is no way around the noise — not for a single qubit and not for more involved quantum evolutions. In Sec. 5.5.2 below we briefly suggest how to put the first law on formal grounds.

5.2. *The second law: Probability distributions described by low-entropy states are noise-stable and can be approximated by low-degree polynomials*

The second law extends our assertions regarding NISQ circuits beyond the NISQ regime. The noise causes the high-degree terms, in a certain Fourier-like expansion of the probability distribution, to be reduced exponentially with the degree. Low-entropy states, for which the effect of the noise is small, have probability distributions expressed by low-degree Fourier terms.[b] Such noise-stable states represent the very low-level computational complexity class, **LDP**, namely, the class of probability distributions that can be approximated by low-degree polynomials. Our second law applies to quantum evolutions in nature that can be described by quantum circuits, and it is a plausible assumption that this applies universally (under some caveats; see Sec. 8.22). We can expect that the specific "Fourier-like expansion" will be different for different physical settings but that the same computational class **LDP** will apply in general.

5.3. *The third law: Entanglement is accompanied by correlated errors*

The third law asserts that the errors for the two qubits of a cat state necessarily have a large positive correlation. Here also we extend well-accepted insights of NISQ systems into general quantum systems. Correlated errors is an observed and accepted phenomenon for gated qubits and, without quantum error-correction, it extends and applies to all pairs of entangled qubits. An important consequence of the third law is that complicated quantum states and evolutions lead to *error synchronization*, namely, to a substantial probability that a large number of qubits, far beyond the average rate of noise, are hit by noise.

We emphasize that the third law is not based on a new way to model noisy quantum circuits, but rather is derived from ordinary models under the assumption that $\beta < \delta$, namely, that the error rate cannot be reduced to the level that enables quantum error-correction. It would be interesting to test the quantitative aspects of the law both by simulation and by experi-

---

[b]For the definition of entropy see Sec. 5.5.1. Meanwhile, we can think of "low entropy" as a synonym for "high fidelity."

ments. See also Sec. 8.21. We note that this law is related to our proposed modeling in Sec. 7.3 (Eq. (11)) but is not related to correlations in the computation of the fidelity via Formula (77), that we discuss, in the context of Google's experiment, in Secs. 6 and 7.

5.4. *The fourth law: Quantum noise accumulates*

The fourth law expresses the fact that without noise cancellation via quantum fault-tolerance, quantum noise must accumulate. In Sec. 5.5.2 we briefly suggest how to put the fourth law on formal grounds.

5.5. *Under the mathematical lens: Noise, time, and non-commutativity*

5.5.1. *Noise, mixed states, density matrices, and entropy*

In quantum physics, states and their evolutions (the way they change over time) are governed by the Schrödinger equation. A solution of the Schrödinger equation can be described as a unitary process on a Hilbert space, and the states (which are called "pure states") are simply unit vectors in this Hilbert space. Quantum computers, as described above, form a large class of such quantum evolutions, and it is even a common view that all quantum processes in nature (or at least all "local" quantum processes) can be described efficiently by quantum computers. When you add noise to the picture you encounter more general types of states (called "mixed states") that can be described (not in a unique way) as a classical probability distribution of pure quantum states.[c] Mathematically speaking, if ρ is a pure state and hence a (row) unit vector in (say) an N-dimensional space, we represent ρ by the matrix $\rho^{tr}\rho$. (This matrix is the outer product of ϕ with itself; in the quantum "bra-ket" notation we write it as $|\rho\rangle\langle\rho|$.) A convex combination of such matrices represents a general mixed state and this representation is referred to as the density matrix representation.[d] The von Neumann entropy $S(\rho)$ of a state ρ (in terms of the density matrix

---

[c]An alternative description of noisy states and evolutions can be given in terms of a larger Hilbert space $H' \supset H$, and a unitary process on H'.

[d]Quantum evolutions on density matrices are described by "quantum operations." We will not discuss them here, but merely mention that their study was the starting point of central areas in mathematics.

description) is defined by $S(\rho) = -tr(\rho \log \rho)$. (Here we refer to logarithm as a function on matrices and logarithm is taken to the base 2.) The entropy is always non-negative and, for a state ρ, $S(\rho) = 0$ if and only if ρ is a pure state.

5.5.2. *Commutativity, time, and time-smoothing*

I will now briefly describe some mathematical ideas required for putting the first and fourth laws on more formal grounds. One obstacle we face when trying to mathematically express the claim that time-dependent evolutions are noisy is that the parameterization of time we start with is arbitrary. We need to consider a canonical parameterization of time. Now, you may recall that two operators U and W (or matrices) are commutative if $UW = WU$. For two operators U and W that do not commute (namely, $UW \neq WU$) a non-commutativity measure refers to a quantitative way to measure by how much U and W fail to commute.

The first law (reformulated): Noise in a certain time interval is bounded from below by a non-commutativity measure of the involved unitary operators.

Furthermore, such a non-commutativity measure can be regarded as an intrinsic parameterization of time for a quantum evolution.

The first law asserts that when you look at a quantum computer that in a certain time interval executes a sequence of unitary operators $U_1, U_2, ..., U_s$, then the amount of noise at that time interval is bounded from below by a non-commutativity measure of those unitary operators. If you start with a single qubit and apply a random sequence of 1-qubit gates you recover the assertion that the quality of a qubit has an absolute positive lower bound. Time dependence allows us to formulate a general law for lower bounds on the amount of noise and to put the intuition that quantum systems are inherently noisy on formal grounds. We note that the first law does not imply that every time-independent quantum evolution can be realized without noise.

We end the section with a brief discussion of the fourth law. The fourth law asserts that quantum noise must accumulate and that noise cancellation via quantum fault-tolerance is not possible. To express this idea mathematically we model "noise accumulation" by considering a subclass of all noisy quantum evolutions where the noise is given by a certain time-smoothing operation.

The fourth law (reformulated): Noisy quantum evolutions are subject to convoluted time-smoothed noise.

Convoluted time-smoothing is a certain mathematical operation that averages out the error over time. (For the definition, see Ref. 9 [Sec. 4.6.2], Ref. 13.) The crucial property is that the "convoluted time-smoothing" can be applied to every noisy quantum evolution, but not every noisy quantum evolution is obtained by such smoothing. We thus end up with a subclass of all noisy quantum evolutions, which is suggested as a class of evolutions where quantum noise necessarily accumulates. We face the difficulty that for general quantum evolutions time parameterization is arbitrary and, here too, we need to take the parameterization of time for the smoothing to be intrinsic.

6. The Google Supremacy Claims

6.1. *The experiment*

The Google experiment is based on the building of a quantum computer (circuit) with n qubits that performs m rounds of computation. The computation is carried out by a 1-qubit and 2-qubit gates. At the end of the computation the qubits are measured, leading to a probability distribution on 0-1 vectors of length n. For the ultimate experiment ($n = 53$, $m = 20$, 1113 1-qubit gates, 530 2-qubit gates) the Google team produced a sample of a few million 0-1 vectors of length 53.

The specific circuit C used for the computation is a random circuit. For every experiment, the specific gates are chosen, once and for all, at random (by a classical computer). Without noise the quantum computer will produce samples from a certain probability distribution D_C that depends on the specific circuit C. Google's quantum computers (like any other quantum computers currently available) are "noisy," so what the computer is actually producing are not samples from D_C but rather a noisy version that can roughly be described as follows: a fraction F of the samples are from D_C and a fraction $(1 - F)$ of the samples are from a uniform distribution. F is referred to as the *fidelity*.

6.2. *The Google supremacy claims*

The paper made two crucial claims regarding the ultimate 53-qubit samples.

A) The fidelity F of their sample is above $1/1000$.

B) Producing a sample with similar fidelity would require 10,000 years on a supercomputer.

6.3. *Google's argument*

As it was only possible to give indirect evidence for both these claims, we shall now describe the logic of Google's quantum supremacy argument.

For claim A) regarding the value of F, the paper describes a statistical estimator for F and the argument relies on a bold extrapolation argument that has two ingredients. One ingredient is a few hundred experiments in the classically tractable regime: the regime where the probability distribution D_C can be computed by a classical computer and the performance of the quantum computer can be tested directly. The other ingredient is a theoretical formula for computing the fidelity. According to the paper, the fidelity of entire circuits closely agrees with the prediction of the simple mathematical formula (Formula (77) in Ref. 14; Eq. (7) below) with a deviation below 10–20 percent. There are around 200 reported experiments in the classically tractable regime including ones carried out on simplified circuits (which are easier to simulate on classical computers). These experiments support the claim that the prediction given by Formula (77) for the fidelity is indeed very robust and applies to the 53-qubit circuit in the supremacy regime. We note that the samples for the 53-qubit experiment demonstrating "supremacy" are archived, but that it is not possible to test them in any direct way.

For claim B) regarding the classical difficulty, the Google team mainly relies on extrapolation from the running time of a specific algorithm they use. They also rely on the computational complexity support for the assertion that the task at hand is asymptotically difficult. (It is also to be noted that using conjectured asymptotic behavior for insights into the behavior in the small and intermediate scales relies on a naturalness assumption.)

6.4. *Estimating the fidelity*

Google's statistic F_{XEB}

Once the quantum computer produces m samples $x_1, x_2, \ldots, x_m$, the following estimator for the fidelity is computed:

$$F_{XEB} = 2^n \frac{1}{m} \sum_{i=1}^{m} D_C(x_i) - 1. \tag{6}$$

Google's *a priori* fidelity prediction

The Google argument relies crucially on the following simple formula (Formula (77) in Ref. 14) for estimating the fidelity F of their experiments:

$$F = \prod_{g \in \mathcal{G}_1}(1 - e_g) \prod_{g \in \mathcal{G}_2}(1 - e_g) \prod_{q \in \mathcal{Q}}(1 - e_q). \tag{7}$$

Here $\mathcal{G}_1$ is the set of 1-gates (gates operating on a single qubit), $\mathcal{G}_2$ is the set of 2-gates (gates operating on two qubits), and $\mathcal{Q}$ is the set of qubits. For a gate g, the term e_g in the formula refers to the fidelity (probability of an error) of the individual gate g. For a qubit q, e_q is the probability of a read-out error when we measure the qubit q. If we replace the detailed individual values for the fidelities by their average value we get a further simplification:

$$F' = (1 - 0.0016)^{|\mathcal{G}_1|}(1 - 0.0062)^{|\mathcal{G}_2|}(1 - 0.038)^n. \tag{8}$$

The rationale for Formula (77) (Eq. (7)) is simple: as long as there are no errors in the performance of all the gates and all the measurements of the qubits, then we get a sample from the correct distribution. A single error in one of these components leads to an irrelevant sample. The Google paper reports that for a large number of experiments the actual fidelity estimated by Formula (77) (Eq. (7)) agrees with the statistical estimator for the fidelity up to 10%–20% percent. We can expect that the value of F' will be a few percentage points higher than that[e] of F. For the circuits

---

[e] An even better approximation is $(1 - 0.0093)^{|\mathcal{G}_2|}(1 - 0.038)^n$.

used by Google, when the number of qubits is n and the number of layers is m (m is an even integer), $|\mathcal{G}_1| = n(m+1)$ and $|\mathcal{G}_2| \leq nm/2$.

Google's statistical philosophy

A basic statistical idea in the Google paper (Ref. 14, [Sec. IV]) is the following:

> Crucially, XEB does not require the reconstruction of experimental output probabilities, which would need an exponential number of measurements for increasing number of qubits. Rather, we use numerical simulations to calculate the likelihood of a set of bitstrings obtained in an experiment according to the ideal expected probabilities.

6.5. *Under the mathematical lens: The Porter–Thomas probability distributions, Archimedes, and size-biased distributions*

What does a "random" probability distribution look like?

Let X be a set and our task will be to describe a "random" probability distribution D on X. Consider another real probability distribution Z where Z is a positive real number and $\mathbb{E}(Z) = 1$. Now, to $x \in X$ we assign a probability $z(x)/|X|$ drawn at random from Z. (To make sure that those are indeed probabilities you need to normalize $\sum_{x \in X} z(x)$ to 1.) This construction was made in a nuclear physics paper by Porter and Thomas (1956)[15] for the case where Z is an χ^2-distribution. The general construction was made in a statistics paper by Kingman (1975).[16]

In the case of Google's experiment, $X = \Omega_n$ (the set of all 0-1 vectors of length n) and Z is the exponential distribution with density function e^{-z}. Also, D is not really random: it is a pseudorandom distribution with properties very similar to those of a truly random distribution. Here, by pseudorandom we mean a value, drawn by a computer program, that behaves "like" a random value. The twist here is that the computer program is a quantum computer program. The assumption behind the quantum advantage claims is that computing this pseudorandom distribution is a very hard problem for a classical computer, yet sampling from this distribution can be easily carried out by a quantum computer.

The exponential distribution, Archimedes, and moment maps

The state of a quantum computer that performs a random sequence of gates is similar to a random unit vector in the Hilbert space described by the computer. Now, when you consider a random unit vector in a high-dimensional complex vector space, the distributions of the real and complex parts of each coordinate are close to Gaussian and, therefore, the distribution of their sum of squares is exponential. Indeed, recall that, in general, the sum of squares of k statistically independent Gaussians is χ_k, the χ-square distribution with k degrees of freedom, and for $k = 2$ this is the exponential distribution. (The statistical independence condition approximately holds for random unit vectors in high dimensions.)

There is a further interesting mathematical story related to why the probabilities $D_C(x)$ behave according to a Porter–Thomas distribution based on an exponential distribution Z. The space Δ of all probability distributions on Ω_n is a simplex of dimension $2^n - 1$. Now, consider a point, drawn at random from a unit sphere in a complex space of dimension 2^d. When we replace "amplitudes" (complex coefficients) by the associated real probabilities, we obtain (precisely, on the nose) a random probability distribution, namely, a point from this simplex Δ drawn uniformly at random. As pointed out by Greg Kuperberg,[17] the connection between the complex amplitudes and the probability distribution is related to a theorem of Archimedes (c. 287–c. 212 BC), whereby a natural projection from the unit sphere to a circumscribing vertical cylinder preserves area. (It is also related to the "moment map" in modern symplectic geometry.)

Statistics: Size-biased distributions

Let us suppose that you want to estimate the distribution D of the number of people in apartments. You sample random people on the street and ask each one how many people share his apartment with him. The distribution, E, of answers will not be identical to D: a quick way to see this is based on the fact that people you meet on the street are not from empty apartments. We face a similar situation when we let the quantum computer sample $x \in \Omega_n$ (this is an analog to the random person we meet on the street) and then compute $D_C(x)$ (this is an analog to asking about how many people share his apartment). The resulting size-biased distribution is given by $\Gamma = xe^{-x}$, and constitutes the basis for the statistical estimator F_{XEB} for the fidelity F. For more on size bias see Refs. 18 and 19.

Google's statistics F_{XEB}
Recall that once the quantum computer produces m samples $x_1, x_2, \ldots, x_m$, the following statistic is computed:

$$F_{XEB} = 2^n \frac{1}{m} \sum_{i=1}^{m} D_C(x_i) - 1.$$

The expected value of $2^n D_C(x)$ when x is drawn uniformly at random is

$$\int_0^{\infty} xe^{-x}dx = 1,$$

while the expected value of $D_C(x)$ when x is drawn from the distribution D_c itself is

$$\int_0^{\infty} x^2 e^{-x}dx = 2.$$

It follows that when x is drawn from the distribution $FD_C + (1 - F)U$, the expected value of $2^n D_c(x)$ is $1 + F$ and, therefore, F_{XEB} is an unbiased estimator for the fidelity F.

7. Preliminary Assessment of the Google Claims

The Google experiment represents a very large leap forward with regard to several aspects of the human ability to control noisy quantum systems. Accepting the Google claims requires a very careful evaluation of the experiments and, of course, successful replications as well. The burden of producing detailed documentation of the experiments and carefully examining the experimental data and that of replications lies primarily with the Google team itself and, naturally, also with the scientific community as a whole.

In my view, there are compelling reasons to doubt the correctness of the Google supremacy claims. Specifically, I find the evidence for the main supremacy claim A) concerning the 53-qubit samples too weak to be convincing.

Furthermore, in my opinion, there are compelling reasons to question the crucial claims regarding perfect proximity between predictions based on the 1- and 2-qubit fidelity and the circuit fidelity. Some of the outcomes reported in the paper appear to be "too good to be true"; that is, the experimental outcomes are unreasonably close to the expectations of the experimentalists. In this section we shall focus on the main example of this type.

It is to be noted that there are also several works that challenge Google's claim B) regarding the complexity of their sampling task on a classical computer. A team from IBM[20] demonstrated a way of improving the running time by 6 orders of magnitude. Another group[21] demonstrated an improvement all the way to within 1–2 orders of magnitude above the quantum running time for a related (albeit easier) sampling problem. Yet another group[22] proposed a tensor network-based classical simulation algorithm for Google's circuit. (See also Sec. 9.)

7.1. *Formula (77): An amazing breakthrough or a smoking gun?*

As you may recall, Formula (77) in the Google paper (Eq. (7), Sec. 6.4) provides an estimation of the fidelity of a circuit based on the fidelities of its components:

Formula (77) $$F = \prod_{g \in G_1}(1-e_g)\prod_{g \in G_2}(1-e_g)\prod_{e \in Q}(1-e_q).$$

The Google paper claims that this formula estimates with a precision of 10%–20% the probability of the failure (fidelity) of a circuit. This remarkable agreement is a major new scientific discovery and it is not needed for building quantum computers. Reaching sufficiently high fidelity levels is indeed crucial, but the demonstration of such accurate predictions on the fidelity based on the error rates of the individual components is neither plausible nor required. The precise fidelity estimation is only needed for the specific extrapolation argument leading to the Google team's supremacy declarations.

In my opinion the claim regarding the fidelity estimation is very implausible and even if quantum computers will eventually be built we are not going to witness the realization of this particular claim. Of course, it might be interesting to check whether we ever see anything remotely like this for other groups attempting to build quantum circuits, or indeed whether we ever see in any other field of engineering such a good estimation of the failure probability of a physical system, with hundreds of interacting elements, as the product of hundreds of individual error probabilities.

The Google team's interpretation of this discovery is that it shows that there is "no additional decoherence physics" when the system scales, and they justify the remarkable predictive power of their Formula (77) (Eq. (7)) with a statistical computation that is based on the following three ingredients:

(1) Individual read-out and gate errors are accurate. The Google team reported that the level of accuracy for the individual qubit and gate fidelities is ±20%.
(2) Errors for the individual fidelity estimates are unbiased; namely, there are no systematic errors.
(3) Error probabilities are statistically independent.[f]

In my view all these claims are questionable and the second and third claims are very implausible. This suggests that the excellent quality of the predictions based on Formula (77) may reflect naive statistical experimental expectations rather than physical reality.

A few remarks: Let me first explain the issue of biased versus unbiased estimation (the second item) with a simplified example. Suppose that you have a space rocket with 900 components and the probability of any component failing is estimated at 0.01. If one component fails, the entire space rocket fails. Under a statistical independence assumption, the probability of success is $(1 - 0.01)^{900}$, which roughly is 0.00012. If your estimate of 0.01 for each individual component is correct up to an unbiased error of 20% (namely, with probability 1/2 the correct error probability is 0.012 and with probability 1/2 it is 0.008), then the deviation of the outcome can be estimated within roughly 3%. But if your estimation is systematically biased in one direction by 20% then the effect on the

---

[f]Based on these assumptions, Google's (rough) estimation of the deviation of the prediction of Formula (77) is

$$0.2 \cdot (\sqrt{n} \cdot 0.038 + \sqrt{|G_1|} \cdot 0.0016 + \sqrt{|G_2|} \cdot 0.0063). \qquad (9)$$

(So, say, for $n = 53$ and $m = 14$ this gives roughly 8.8%.) However, The gap between the (77) prediction and the fidelity estimation based on the data, while bounded at 10%–20%, does not increase with n as Formula (9) suggests.

probability of success is by a factor of five or so.

We also note that positive correlation between the error probabilities will actually lead to higher fidelity. There is, in fact, an entire discipline, in statistics and systems engineering, called reliability theory, that studies failure properties of devices based on the failure distributions of individual components.

Finally, an explanation for the success of Formula (77), suggested by Peter Shor (in a discussion in my blog) and various other scholars,[23] is that the statistical independence needed for the success of Formula (77) is justified for random circuits. I do not see a justification for this claim, but it surely deserves further study.

7.2. *What needs to be done*

Listed below are steps required for a further assessment of the Google supremacy claims:

- Further documentation of past experiments and a more careful documentation of future experiments.
- Replications of the experiments by the Google team: larger samples and further experiments in the classically tractable regime; further experiments in the 40–53 qubit range.
- Blind tests: some of the required replications by the Google team should apply the standard methodology of blind tests.
- Replications by other groups of various aspects of the Google claims, including the supremacy claims, the fidelity prediction claims, and the calibration methodology.[g]
- Careful examination of the supremacy experiments both by the Google quantum-computing group itself, by the scientific community, and by Google.

---

[g]Here, I mean "replications" in a broad sense: replications by other groups need not apply the precise Google 2-qubit coupler. We can learn a lot from sampling based on a random circuit with standard 2-qubit gates, and if doing it for 53 qubits is too difficult, reliable experiments on 20–30 qubits could already be useful. A clear challenge would be to replicate (even in these easier settings) the prediction power of formula (77), or even something only ten times worse.

7.3. *Under the mathematical lens: Noise, variance, and Pythagoras*

Another aspect of the experiment that deserves thorough examination is the extent to which the noisy distributions presented by Google's experiment fit the theoretical expectation. This is one aspect of the work I am currently conducting with Yosi Rinott and Tomer Shoham.[19] In this section we talk about several interesting mathematical and statistical aspects of distributions produced by NISQ circuits.

A toy model for the noise of quantum circuits

Below is a simple toy model of what the noisy version of a quantum sampling problem may look like. It is based on the model from Sec. 4.4. Let $D(x_1, x_2, \ldots, x_n)$ be a probability distribution on 0-1 vectors of length n. Given a parameter t we consider the noisy version of D as

$$N_t(D)(x) = \sum_{y \in \Omega_n} D(x+y) t^k (1-t)^{n-k}. \tag{10}$$

Here, again, $y = (y_1, y_2, \ldots, y_n)$ is also a 0-1 vector and $y_i = 1$ indicates "error in the ith coordinate." The sum $x + y$ should be considered as a sum modulo 2: $x_i + 0 = x_i$ and $x_i + 1 = 1 - x_i$. If E is a probability distribution on Ω_n then we can consider a more general form of noise, namely,

$$N_t(D)(x) = \sum_{y \in \Omega_n} D(x+y) E(y). \tag{11}$$

Equation (10) is the case where $E(z) = B_t(z) = t^k (1-t)^{n-k}$, where $k = |z|$. For random (or pseudorandom) quantum circuits, I expect that the effect of the noise on gates will be close to our model for the case where E is a mixture of $B_t(y)$'s (more specifically, a Curie–Weiss distribution), and that this mixture will have a strong positive correlation between errors. Modeling the noise by equations (10, 11) abstracts away the dependence of noise on the structure of the circuits and I expect that such modeling will be useful both qualitatively and quantitatively.

The second-order term of noise

Let us now move from an abstract study of noise to the Google experiment. A simple approximation of the noisy distribution considered by Google is

$$FD_C + (1-F)U, \tag{12}$$

where F is the fidelity. Namely, with probability F we sample according to D_C and with probability $(1 - F)$ we sample according to the uniform probability distribution.

A more detailed description that we may expect is of the form

$$FD_C + (1 - F)N_C, \tag{13}$$

where N_C is a small fluctuation of the uniform distribution that also depends on the circuit C. As it turns out, this more detailed form of noise does not affect Google's size-biased distribution and the F_{EXB} estimator for the fidelity. Yet such detailed descriptions of the noise can be examined by performing similar tests specifically geared to the noise N_C.

Let us denote by F_g the probability that no error occurs for 1-qubit or 2-qubit gates. We can split the noisy distribution into three parts,

$$FD_C + (F_g - F)N_{RO} + (1 - F_g)N_G, \tag{14}$$

where N_G describes errors that involve also faulty gates, and N_{RO} describes the effect of read-out errors when there are no faulty gates. For the read-out errors, Eq. (10) appears to give a good approximation, particularly under Google's statistical independence assumption of read-out errors. Let e_i denote the error probability for the ith qubit; then,

$$(F_g - F)N_{RO} = (F_g - F) \sum_{y \in \Omega_n, y \neq 0} D_C(x + y) \prod_{i:y_i=1} (e_i) \prod_{i:y_i=0} (1 - e_i). \tag{15}$$

If we use averaged errors as in Eq. (8) we reach a simpler formula. Let $F' = (1 - 0.0016)^{|\mathcal{G}_1|}(1 - 0.0063)^{|\mathcal{G}_2|}(1 - 0.036)^n$, and $F'_g = (1 - 0.0016)^{|\mathcal{G}_1|}(1 - 0.0063)^{|\mathcal{G}_2|}$. We replace $F'D_C + (1 - F')U$ with $F'D_C + (F'_g - F')N'_{RO} + (1 - F'_g)U$ with

$$(F'_g - F')N'_{RO} = (F'_g - F') \sum_{y \in \Omega_n, y \neq 0} D_C(x + y)(1 - 0.036)^{|y|}(0.036)^{n - |y|}. \tag{16}$$

Variance computation and Pythagoras

Let me refer to a problem that was raised in relation to the variance estimation of this statistical parameter. Given a circuit C one can estimate the variance of the parameter for various samples. However, when considering the required size of samples for several experiments for various circuits, one needs to compute the variance across different circuits, while using the following formula:

$$var(A) = \mathbb{E}(var(A|B)) + var(\mathbb{E}(A|B)). \qquad (17)$$

When my friend and colleague Yosi Rinott teaches this formula for computing the variance, he tells the students that they have surely seen this formula before. For us it is an opportunity to see Greg Kuperberg's reference to Archimedes (Sec. 6.5) and raise him another 200 years (backwards) to Pythagoras (c. 570–c. 495 BC). Indeed, Eq. (17) is just a disguised form of the Pythagorean theorem.

A glimpse into my study with Yosi Rinott and Tomer Shoham

1) Our study[19] of the fidelity estimate of the Google team confirms that a more precise description of the noise (of the kind considered above) will not make a difference in the expected value of F_{XEB} and will make only a small insignificant difference in the variance. (Here the Pythagorean formula for the variance (Eq. (17)) comes into play.) (In general, both F_{XEB} and the entire size-biased empirical distribution are fairly robust.) It also confirms and extends results of the Google team asserting that compared to other (moment) estimators of a similar nature, F_{XEB} has smaller variance and therefore smaller samples are required for definite results.

2) Google's samples would allow us to check on the data our proposal for the read-out noise, N_{RO}. This provides an alternative fidelity estimator allowing to test the quality of the data of the Google experiment. (So far, this alternative fidelity estimation has been checked only for $n = 12, 14$.)

3) A preliminary study of the Google data on 12 and 14 qubits further suggests that neither Google's basic noise model nor our refined read-out model fits the observed data, and the second moment of the empirical distribution is considerably higher than what the models predict. On the other hand, there is a perfect agreement between experiment and theory regarding the size-biased distribution (Fig. S32 in Ref. 14) that also deserves examination. For 12 and 14 qubits the data also exhibits non-stationary behavior (that might be chaotic). This seems consistent with the noise-sensitivity pictures from Secs. 4.3 and 4.4 and deserves to be examined for other NISQ samples.

4) Another finding from Ref. 19 that is also related to the the analysis in Ref. 1 (Sec. IV.A, especially Formulas (17,21)), is the following: when

one considers a probability distribution based on a *specific* realization of a Porter–Thomas distribution, then the Google statistic F_{XEB} is no longer an unbiased estimator. We asserted that when x is drawn from the distribution $FD_C + (1 - F)U$, the expected value of $2^n D_C(x)$ is $1 + F$ and, therefore, F_{XEB} is an unbiased estimator for the fidelity F. This assertion is correct over all realizations of the Porter–Thomas distribution (or over all random circuits C), but for a specific realization (or a specific circuit C), F_{XEB} is biased. The expected value of $2^n D_C(x)$ is $1 + \alpha F$, where

$$\alpha = -1 + 2^n \sum (D_C(x))^2. \tag{18}$$

This leads to a similar yet better estimator (referred to as V) for the fidelity that depends on the specific circuit C. In Ref. 19 we also study the maximum likelihood estimator (MLE) which is superior to other estimators mentioned here (and is also unbiased for every realization). These observations suggest an interesting improvement of Google's main statistical tool (when the number of qubits is not large).

8. Possible Connections and Applications

In this section we mention various potential applications and connections to physics arising from a fundamental failure of quantum computation and quantum error-correction. Also here, the proposed connections and applications largely rely on the argument against quantum computers and a fundamental failure of quantum computation and quantum error-correction. Yet, a few of the insights described in this section can apply to fragments of quantum physics and quantum engineering even in the case where quantum computers are possible. Finally, we explore also strange counterintuitive consequences of the reality without quantum computation that may even weaken the argument against quantum computers.

8.1. *Time and geometry*

For classical computers, the program you run is not restricted by the geometry of the computer, and the information described by a piece of your hard disc does not depend on the geometry of that piece. This is such an obvious insight that we do not even spare it a second thought. Universal quantum computers will allow implementing quantum states and quantum evolutions on an array of qubits of arbitrary shape. On the other hand, the impossibility of quantum error-correction suggests that quantum states and evolutions constrain the geometry. The failure of quantum fault-tolerance will contradict computer-based intuitions that the information does not restrict the geometry, but will agree with insights from physics, where witnessing different geometries supporting the same physics is unusual and important. An example of an important geometric distinction, when it comes to quantum behavior, is the different behavior of physics of different geometric scales: we witness very different microscopic physics, mesoscopic physics, and macroscopic physics.

The same is true for time. With quantum fault-tolerance, every quantum evolution that can experimentally be created can be time-reversed and, in fact, we can permute the sequence of unitary operators describing the evolution in an arbitrary way. In a reality where quantum fault-tolerance is impossible, time reversal is not always possible

It is a familiar idea that since (noiseless) quantum systems are time-reversible, time emerges from quantum noise (decoherence). (This idea has its early roots in classical thermodynamics.) Putting geometry and time together, we can propose that, generally speaking, quantum noise and the absence of quantum fault-tolerance enable the emergence of time and geometry.

8.2. *Superposition and teleportation*

In a recent paper about the future of physics, Frank Wilczek (2015)[24] predicts that large-scale quantum computers will eventually be built and describes why these excite him: "A quantum mind could experience a superposition of 'mutually contradictory' states [...] such a mind could revisit the past at will, and could be equipped to superpose past and present. To me, a more inspiring prospect than factoring large numbers."

Indeed, superposition is at the heart of quantum physics — and a common intuition that is supported by an ability to build universal quantum computers is that for every two quantum states that can be constructed, their superposition can also be constructed. Similarly, a common intuition is that every quantum state that can be prepared can also be teleported.

A central insight stemming from the argument against quantum computing (and the various proposed laws associated with it) is that already for a small number of qubits certain pure states cannot be well approximated. (The fidelity F is a good measure for what "well approximated" means.) For two pure states ρ_1, ρ_2 that can be achieved but are close to the limit, a superposition between ρ_1 and ρ_2 that requires a more complicated circuit than that needed for ρ_1 and ρ_2 may already be beyond reach. By the same token, there is a quantum state ρ that can be well approximated but is close to the limit, and cannot be teleported. The reason is that a circuit needed to demonstrate a teleportation for ρ is considerably more involved than a circuit needed to demonstrate ρ.

8.3. *Predictability and chaos*

Noise sensitivity asserts that for very general situations the effect of the noise will be devastating. This means that the actual outcomes not only will largely deviate from the ideal (noiseless) outcomes but also will be very dependent on fine parameters of the noise, thus leading to processes with large chaotic components.

8.4. *The black-hole information paradox*

Quantum information and computation play a role in explanations of the black-hole information paradox.[h] Of particular importance in these explanations are "pseudorandom" quantum states of the kind Google attempts to build (but on a much larger number of qubits). According to our laws, such pseudorandom quantum states cannot be achieved locally, and this goes against the rationale of some of the attempted solutions. On the other hand, our laws asserting that A) qubits are inherently noisy and B) entanglement is necessarily accompanied by correlated noise may already suggest a resolution to some versions of the "paradox" (e.g., to those based on no-cloning or on monogamy of entanglement).

---

[h]In the absence of a definite theory of quantum gravity, the paradox can be seen as lying between the foundation of physics and philosophy.

8.5. *The time-energy uncertainty principle*

The time–energy uncertainty principle (TEUP) is a much-studied (controversial) issue in quantum mechanics. Counterexamples were given by (Yakir) Aharonov and Bohm,[25] and are based on the ability to prescribe time-dependent quantum processes. A counterexample to an even weaker and more formal version of TEUP was given by (Dorit) Aharonov and Atia[26] based on Shor's factoring algorithm. Our study casts doubt on the very ability to prescribe noiseless time-dependent quantum evolutions at will, while also challenging the feasibility of Shor's algorithm, and thus the picture drawn here in fact militates against the physical relevance of these counterexamples.

8.6. *Realistic models for fluctuations*

One interesting property suggested by a critical look at the theory of quantum fault-tolerance is that fluctuations in quantum systems with an (even small) amount of interaction are super-Gaussian (perhaps even linear). Here, we challenge one of the consequences of the general Hamiltonian models allowing quantum fault-tolerance (see, e.g., Ref. 27). These models allow for some noise correlation over time and space but they are characterized by the fact that the error fluctuations are sub-Gaussian. Namely, when there are N qubits the standard deviation for the number of qubit errors behaves like $\sqrt{N}$ and the probability of more than $t\sqrt{N}$ errors decays as it does for Gaussian distributions.

There are various quantum systems where the study of fluctuations will prove interesting. For example, systems for highly precise physical clocks are characterized by having a huge number N of elements with extremely weak interactions. We still expect (and this may even be supported by current knowledge) that in addition to $\sqrt{N}$-fluctuations there will also be some εN-fluctuations. Of course, the relation between the level of interaction and ε is of great interest. (The intuition of sub-Gaussian fluctuations may even be more remote from reality for engineering devices and this is also related to our discussion of Google's Formula (77).)

8.7. *The unsharpness principle*

The unsharpness principle is a property of noisy quantum systems that can be proved for certain quantizations of symplectic spaces. This was studied by Polterovich (in Ref. 28) who relies on deep notions and results from symplectic geometry and follows, on the quantum side, some earlier works by Ozawa[29] and Busch, Heinonen, and Lahti.[30] Here, the crucial distinction is between general positive operator-valued measures (POVMs) and von-Neumann observables, which are special cases of POVMs (also known as projector-valued POVMs). The unsharpness principle asserts that (under some locality condition) certain noisy quantum evolutions described by POVMs must be unsharp, namely, "far" from von Neumann observables. The amount of unsharpness is bounded from below by some non-commutativity measure. It is interesting to explore the (mathematical and physical) scope of the unsharpness principle and its connection to our first law.

8.8. *Topological quantum computing*

Topological quantum computing is an approach whereby robust qubits are created not by implementing quantum error-correction on NISQ circuits but by realizing stable qubits via anyons. The argument from Sec. 4 can be extended also to this case (see Ref. 8 [Sec. 3.5]. In any case, it is plausible that topological quantum computing and circuit-based quantum computing will meet the same fate. (See also Sec. 9.)

8.9. *Are neutrinos Majorana fermions?*

Majorana fermions are a type of fermions constructed mathematically by Majorana in 1937 but so far not definitely detected in nature. However, there is a compelling argument that neutrinos (or, more precisely, an expected yet undiscovered heavy type of neutrinos) are Majorana fermions.

At the ICA workshop in Singapore, David Gross commented that anyonic qubits required for topological quantum computing are based on condensed-matter analogs of Majorana fermions, which constitutes a strong argument that anyonic qubits are feasible. Taking this analogy for granted, we can ask whether an argument against topological quantum computing casts doubts on the common (conjectural) expectations for Majorana fermions. However, a review of the literature (e.g., Ref. 31) and consultations with colleagues revealed that Majorana fermions from high-energy physics are most commonly regarded as analogs of more mundane objects (Bogoliubov quasiparticles) from condensed-matter physics. Therefore, the argument against topological quantum computers and stable anyonic qubits does not shed light on the nature of neutrinos (but this is indeed the *kind* of insight we would *hope* to get).

8.10. *Noise stability and high-energy physics*

Extending the framework of noise stability and sensitivity to mathematical objects of high-energy physics is an appealing challenge. Let us assume for a minute that this can be done. We can ask if our second law asserting that realistic quantum states and evolutions are noise-stable provides some insights into the various mysteries surrounding definite, but unexplained, features of the standard model.

8.11. *Does nature support supersymmetry?*

Supersymmetry is a famous mathematical extension of the mathematics of the standard model. It is widely believed that supersymmetry and, in particular, supersymmetric extensions of the standard model are crucial to understanding physics beyond the standard model and quantum gravity. So far, there is no definite experimental support for this belief.

Our second law imposes a severe limitation on quantum states and evolutions and asserts that they can be described within a very restrictive computational class **LDP** of low-degree polynomials. We asked above whether this law can contribute to the understanding of the standard model, and we can ask the same question with reference to the proposed supersymmetric extensions of the standard model. Our second law supports classical error-correction and classical computation but not quantum error-correction and quantum computation, and an appealing analogy might be that the second law does not support supersymmetric extensions of the standard model at all.

8.12. *Cooling and exotic states of matter*

Noise stability, or the bounded-depth/low-degree polynomial description, may shed (pessimistic) light on the feasibility of various exotic states of matter. In some cases, such exotic states of matter are beyond reach, and, in other cases, the computational restriction may apply only to low-temperature states. (As the entropy increases, there are more opportunities to represent our state as a mixture of pure states that abide by the complexity requirement.) Within a symmetry class of quantum states (or for classes of states defined in a different way), noise stability, or the low-degree polynomial description, may provide an absolute lower bound for cooling. An appealing formulation would be that for a class of quantum states the "absolute zero" temperature may depend on the class.

8.13. *The emergence of classical information and computation*

It is an interesting question to find fundamental reasons for why quantum information is more fragile than classical information, see Ref. 32. We propose the following answer: the class **LDP** of functions and probability distributions that can be approximated by low-degree polynomials does not support quantum advantage and quantum error-correction, yet it still supports robust classical information, and with it also classical communication and computation. The "majority" Boolean function has excellent low-degree approximations and allows for very robust classical bits based on a large number of noisy bits (or qubits). It is possible that every form of robust information, communication, and computation in nature is based on classical error-correction where information is encoded by repetition (or simple variants of repetition) and decoded in turn by some variant of the majority function. (On top of this rudimentary form of classical error-correction, we sometimes witness more sophisticated forms of classical error-correction.)

8.14. *Learnability of physical systems*

The theory of computing studies not only efficient computing but also efficient learning, namely, the ability to efficiently learn a member in a class from examples. One major insight is that compared to carrying out

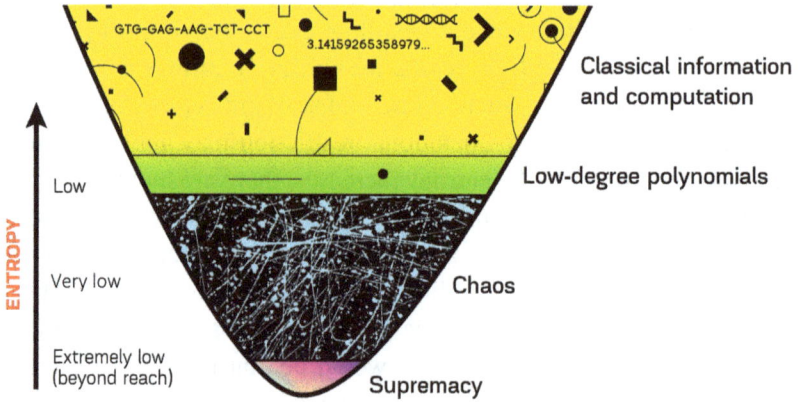

Fig. 2.   Low-entropy quantum states give probability distributions described by low de-gree polynomials, and very low-entropy quantum states give chaotic behavior. Higher entropy enables classical information.

computation when the model is known, it is notably much harder to learn an unknown model. Efficient learning is very restrictive, but our very low-level class **LDP** allows for efficient learning. We note that $\mathbf{AC^0}$ which is a larger class and yet a very low level one, does *not* allow, in general, efficient learning. I don't know if there are approximately learnable dis-tributions beyond $\mathbf{AC^0}$. (See Fig. 1.) Efficient learnability of low-entropy quantum systems may provide an explanation for our ability to understand natural processes and the parameters defining them.

A bold conjecture is that, in practice, robust distributions arising from NISQ systems are practically learnable via standard machine-learning methods.

8.15. *Reaching ground states*

Reaching ground states is computationally hard (**NP**-hard) for classical systems, and even harder for quantum systems. So how does nature reach ground states so often? Quantum evolutions and states approximated by low-degree polynomials represent severe computational restrictions that can make reaching ground states computationally easy, and this provides a theoretical support as to why, in many cases, nature easily reaches ground states.

8.16. *Noise and symmetry*

One insight from the failure of quantum error-correction and the accumulation of noise is that noisy quantum states and evolutions are subject to noise that respects their symmetries.

An interesting example is that of Bose–Einstein condensation. For a Bose–Einstein state on a bunch of atoms, one type of noise corresponds to an independent noise for the individual atoms. Another type of noise represents fluctuations of the collective Bose–Einstein state itself. This is the noise that respects the internal symmetries of the state and it is expected that such a form of noise must always be present.

8.17. *Does Onsager's thermodynamic principle apply to quantum systems?*

(This connection was suggested by Robert Alicki years ago.) Onsager's thermodynamical law expresses the idea that the statistical laws for the noise are related to the statistical laws for the "signal." This idea is related to the effects of noise accumulation and to some of the items previously discussed. There is some controversy regarding the question of whether and how Onsager's law extends to quantum physics and it will be interesting to see whether the proposed counterexamples are in conflict with our restrictions on noisy quantum processes.

8.18. *The extended Church–Turing thesis*

The extended Church–Turing thesis (ECCT) (see, e.g., Refs. 33 and 34) asserts that every realistic computing device can only perform efficient classical computation. Universal quantum computers violate the extended Church–Turing thesis. By contrast, our theory supports the validity of the extended Church–Turing thesis. See Ref. 8 for a detailed discussion. (We note that our theory is not based on the ECCT, but rather on computational complexity considerations for very low-level complexity classes.)

8.19. *Naturalness revisited*

Here are three examples of similar deductions based on the naturalness heuristic (Sec. 2.2) for computational complexity.

The first example is an important part of the theoretical foundation of the Google experiment (See Refs. 1 and 35).

A1) Finding a sample with $F_{XEB} > \varepsilon$ is exponentially hard as a function of n (for a fixed ε).

A2) This supports the assertion that achieving this task (for $\varepsilon = 1/1000$) on 53 qubits represents quantum advantage.

The second example refers to a recent proposal for implementing Shor's factoring algorithm using classical devices called stochastic magnetic circuits.[36]

B1) The computational power of the stochastic magnetic circuits offered for implementing Shor's algorithm is within **P**.

B2) This supports the assertion that these devices offer no superior way to factor integers.

And, finally, the third example is the crux of my argument against quantum computers.

C1) The computational power of NISQ computers is **P** (for a fixed rate, ε, of noise).

C2) This supports the assertion that NISQ computers offer no superior computation.

The naturalness heuristic plays (often in an implicit way) a central role in the way computational complexity insights are related to computational reality. It is relevant to computational complexity insights in practical algorithms, in scientific computing, in practical areas of cryptography, and in machine learning and statistics. This is an interesting topic for further study.

8.20. *"So what about the energy levels of the lithium atom?"*

The argument against superior quantum computation suggests that robust computations performed by nature can, at least in principle, be carried out efficiently on a digital computer. Yet, there are robust physical quantities that "nature computes" for which efficient classical computations (and especially computations "from first principles") are currently unavailable. (For more on this issue, see Ref. 37 [Secs. 6.5 and 4] or Ref. 13.)

8.21. *Correlation and modeling the noise*

A critique of the third law reads as follows:

"Entanglement is a feature of a state (in Hilbert space), not of the operator that acts on the state. The noise is due to which operator acts on the state. In quantum error-correction and fault-tolerance theory we analyze the structure of the operator that acts on the state and show that the locality of the interactions in this operator and the weakness of the unwanted interactions enable fault-tolerance, not whether it [the operator] acts on entangled states or product states. Locality of interactions here means that we have no 10-body interactions, etc.: really every Hamiltonian, field theory, or theory that is ever used in physics is in accordance with this notion, so deviating from this concept seems ill-advised and badly motivated."

This point deserves to be explained: nothing in the theory described here is based on non-local modeling of the noise. As a matter of fact, it is based on the very standard modeling of noisy quantum circuits. Our argument (Sec. 4) asserts that $\beta < \delta$ and therefore quantum error-correction is not possible. Now, it is well accepted both as part of the theory and as an empirical fact that when we create entanglement for two qubits directly by a gate we face correlated errors: depolarizing noise that collapses the state to the maximal entropy state for the four-dimensional Hilbert space describing the pair of qubits. What the third law simply says is that in the absence of quantum error-correction the accumulated errors will be correlated (provided they are still small enough) also for entanglement created indirectly.

8.22. *"It from qubit": Does entanglement explain geometry and gravity?*

Over the past decade, there have been several proposals (often referred to as "it from qubit") that gravity (and other parts of physics) can be understood from insights and techniques derived from quantum information theory and particularly entanglement. People have raised questions like: Does spacetime emerge from entanglement? Can entanglement shed light on gravity? And can quantum computers simulate all physical phenomena?

The idea that spacetime emerges from entanglement is in line with the concept that quantum states restrict time and geometry. Yet, the type of entanglement presented in some of these works is often well beyond the reach of local quantum processes according to our viewpoint. Some proposed connections between spacetime and entanglement might be consistent with a (speculative) possibility that nature is described by more than one local system when certain states that are mundane for one local system are highly entangled for other systems.

8.23. *Theory, reality, and practice*

Many of the items listed in this section may lead to interesting mathematics, and I hope to put some of them under the mathematical lens or, better yet, to see this done by others. Let me suggest a wider context for the discussion, one that encompasses understanding the relation between theory, reality, and practice in computer science, in physics, and in other applications of mathematics.[i]

9. Developments in Recent Months (Written: March 2021)

9.1. *The photonic advantage claim*

A recent paper[2] published in *Science* claims to achieve "quantum computational advantage" at room-temperature using photons. Specifically,

---

[i]The relations between the theory of computing and practical reality was one of the themes in my ICM2018 paper,[9] and it is based on three examples: linear programming, voting methods, and quantum computers.

the paper reports a Gaussian boson sampling experiment representing a quantum state in $\sim 10^{30}$-dimensional Hilbert space and a sampling rate that is $\sim 10^{14}$ faster than that of using digital supercomputers. This paper was described as the first independent verification of the Google's quantum advantage claims. In fact, the claimed advantage is several orders of magnitude higher than Google's claims.

This huge computational advantage claim is based on certain statistical tests measuring the proximity of the empirical samples to the outcomes of noiseless simulations of the quantum experiment. (The simulations were run on a digital supercomputer.) The statistical reasoning of Ref. 2) is based on comparing the empirical samples to a few other distributions. However, in view of the 2014 results of Kalai and Kindler[3] this statistical reasoning is incorrect and therefore the conclusion of achieving huge quantum computational advantage is unfounded. Moreover, a polynomial-time algorithm from Kalai and Kindler[3] may achieve similar or better sampling quality for the statistical methods of Ref. 2. See also Ref. 38 and Renema[39] (and papers cited there). Kalai and Kindler's analysis[3,38] is based on taking a truncated Fourier–Hermite expansion on the boson sampling distribution. Renema's paper proposes another method of "spoofing," namely, another efficient algorithm for achieving similar sampling quality based on an algorithm of Clifford and Clifford.[40]

9.2. *Quantum advantage via quantum annealing?*

Recent claims[41] by scientists from D-Wave and Google have asserted that quantum annealing algorithms performed by a D-Wave quantum computer on several problems are several orders of magnitude faster compared to certain classical packages for the same problems. These claims add to earlier claims by D-Wave scientists from 2014 and 2018. Since, in this case, there is no theoretical foundation for quantum advantage, D-Wave claims have been received with skepticism by the quantum computing community.

9.3. *Are stable non-abelian anyons possible?*

Topological quantum computing is based on creating protected qubits via anyons. As we already mentioned, my argument can be extended to imply that protected topological qubits are not possible either (Ref. 8 [Sec. 3.5]). Now, an important step toward creating protected topological qubits is the creation of certain condensed-matter quasi-particles, and this was claimed in a 2018 paper[42] by researchers from Microsoft. While my argument directly contradicts the huge computational advantage claims, the situation here is more nuanced: I do not know if the quantum states claimed in Ref. 42 can lead to sampling that demonstrates a quantum advantage, or are in conflict with some other laws proposed here. This question may depend on the level of noise, and certainly deserves further study. Be that as it may be, the authors of Ref. 42 have retracted[43] their claims because of inconsistencies between the raw measurement data and the figures that were published in the paper.

9.4. *"Spoofing" Sycamore*

Pan and Zhang[44] proposed a general tensor network method for simulating quantum circuits. As an application, they studied the sampling problem of Google's Sycamore circuits, and announced that by using a moderate computing power they could generate one million (very) correlated bitstrings from the Sycamore circuit with 53 qubits and 20 cycles, with (XEB) fidelity equal to 0.739, which is much higher than the fidelity in Google's quantum supremacy experiments. This result on its own may shed serious doubts on Google's supremacy claims.

9.5. *Under the mathematical lens: Sampling and matching*

All four items described above are related to fascinating mathematics. I will mention only one simple connection that fits in nicely with several topics discussed in this paper. Consider a bipartite graph with a set A of n vertices on one side and a set B of m vertices on the other side, $m \geq n$. Suppose that you want to sample a multi-subset C of B according to the number $n(A, C)$ of semi-matchings from A to C. Here, a multi-subset is a list of elements from B with repetitions, and if C is a multi-subset of B, a semi-matching is a map from A to C such that every vertex is mapped to a neighbor, and the images are precisely the vertices in C with the prescribed multiplicities (see, e.g., Ref. 45). When C is an ordinary subset this is the ordinary notion of matching (see Sec. 2.4). Now, computing $n(A, C)$

is computationally hard (**#P-complete**). However, there is a simple efficient algorithm for the sampling task. (The sampling is on the nose.) The algorithm is based on Ref. 40 and I learned it from Jelmer Renema. Simply, choose for each vertex of A a neighbor in B uniformly at random! This small note gave us an opportunity to consider again matchings that are most fascinating mathematical objects, with great importance in theoretical computer science, and to be reminded that sampling is easier (and can be much easier in some cases) than computing the probabilities. (The computational gap between sampling and computing individual probabilities was already a main insight of Troyansky and Tishby.[12])

10. Conclusion

The crux of the argument against quantum computation is simple. For fixed constant error rates, quantum circuits in the intermediate scale are primitive computational devices. They represent computation in **P** and, more than that, a computational complexity class **LDP** that even allows polynomial-time learnability. This implies that a huge quantum computational advantage is beyond reach for NISQ computers, and therefore the harder task of creating good-quality quantum error-correcting codes is beyond reach as well. This argument is robust since for a large range of sub-constant error rates we can expect chaotic (noise sensitive) behavior. The argument presented here explains why classical computation is possible: the computational class **LDP** supports rudimentary forms of classical error-corrections, and therefore, in the large scale, also robust classical information and computation. The argument applies to all the different proposals for implementing NISQ computers and extends to forms of quantum computation that are not based on NISQ devices such as topological quantum computing. My argument predicts that the recent claims of a huge quantum computational advantage are false. Preliminary work by others and myself supports this prediction.

My work on quantum computation started in 2005 and is marked by three major stages. Until 2013 I mainly studied correlations of errors (for entangled states) and my efforts could be described (in hindsight) as mainly trying to draw conclusions from the failure of quantum fault-tolerance. Some of those conclusions are described in Secs. 5 and 8. The connection to noise stability and noise sensitivity, leading to my computational theoretic argument against quantum computers arose from my 2014 work with Guy Kindler on boson sampling. Conducting a large part of the discussion in English, while at times placing some fragments under the mathematical lens, is characteristic not only of this paper but of my work as a whole.

As of the end of 2019, my argument against quantum computers was challenged by a bold far-reaching experimental claim. Seeking to critically study and possibly refute the Google claims is different from merely seeking to understand the laws of abstract noisy quantum systems. Having an opportunity to rethink matters of statistics (with my colleagues Yosi Rinott, Tomer Shoham, and others) is pleasant, but, on the other hand, trying to understand what is really going on in the Google experiment is also, in various ways, less uplifting. Yet, I also find this pursuit to be of interest and importance that extends beyond the specific case in question. I wish to stress that my critique of the Google experiment was first brought to the attention of the Google team and discussed with them. In the skepticism and debate that have swirled around quantum computing and that I have been involved with in the past 15 years, winning has not been the only thing; indeed, it has not even been the most important thing. What I find important is making the right choices and right judgments in delicate scientific and social situations that are full of uncertainties.

Over the past four decades, the very idea of quantum computation has led to many advances in several areas of physics, engineering, computer science, and mathematics. I expect that the most important application will eventually be the understanding of the impossibility of quantum error-correction and quantum computation. Overall, the debate over quantum computing is a fascinating one, and I can see a clear silver lining: major advances in human ability to simulate quantum physics and quantum chemistry are expected to emerge if quantum computational advantage can be demonstrated and quantum computers can be built, but also if quantum computational advantage cannot be demonstrated and quantum computers cannot be built.

Some of the insights and methods characteristic of the area of quantum computation might be useful for classical computation of realistic quantum systems — which is, apparently, what nature does.

References

1. F. Arute *et al.*, Quantum supremacy using a programmable superconducting processor, *Nature* **574**, 505–510 (2019).
2. H.-S. Zhong *et al.*, Quantum computational advantage using photons, *Science* **370**, 1460–1463 (2020).
3. G. Kalai and G. Kindler, Gaussian noise sensitivity and BosonSampling (2014), arXiv:1409.3093.
4. D. Aharonov and M. Ben-Or, Fault-tolerant quantum computation with constant error, in STOC '97, ACM, New York, 1999, pp. 176–188.
5. A. Y. Kitaev, Quantum error correction with imperfect gates, in *Quantum Communication, Computing, and Measurement* (Plenum Press, New York, 1997), pp. 181–188.
6. E. Knill, R. Laflamme, and W. H. Zurek, Resilient quantum computation: Error models and thresholds, *Proceedings of the Royal Society of London A* **454**, 365–384 (1998).
7. S. Aaronson, and A. Arkhipov, The computational complexity of linear optics, *Theory of Computing* **4**, 143–252 (2013).
8. G. Kalai, The argument against quantum computers, in: M. Hemmo and O. Shenker (eds.), *Quantum, Probability, Logic: Itamar Pitowsky's Work and Influence* (Springer, 2020), pp. 399–422, arXiv:1908.02499.
9. G. Kalai, Three puzzles on mathematics, computation and games, in *Proceedings of the International Congress of Mathematicians* 2018, Rio de Janeiro, Vol. I (2018), pp. 551–606.
10. I. Benjamini, G. Kalai, and O. Schramm, Noise sensitivity of Boolean functions and applications to percolation, *Publications Mathématiques de l'Institut des Hautes Études Scientifiques* **90**, 5–43 (1999).
11. J. Kahn, G. Kalai, and N. Linial, The influence of variables on Boolean functions, in *Proceedings of the 29th Annual Symposium on Foundations of Computer Science*, 1988, pp. 68–80.
12. L. Troyansky and N. Tishby, Permanent uncertainty: On the quantum evaluation of the determinant and the permanent of a matrix, in *Proceedings of the 4th Workshop on Physics and Computation*, 1996.
13. G. Kalai, The quantum computer puzzle, *Notices of the American Mathematical Society* **63**, 508–516 (2016).
14. F. Arute *et al.*, Supplementary information for "Quantum supremacy using a programmable superconducting processor" (2019), arXiv:1910.11333.
15. C. E. Porter and R. G. Thomas, Fluctuations of nuclear reaction widths, *Physical Reviews* **104**, 483–491 (1956).
16. J. F. C. Kingman, Random discrete distributions, *Journal of the Royal Statistical Society, Series B* **37**, 1–22 (1975).
17. G. Kuperberg, Archimedes' other principle and quantum supremacy, Guest post on "Shtetl Optimized," Nov. 2019.
18. R. Arratia, L. Goldstein, and F. Kochman, Size bias for one and all, *Probability Surveys* **16**, 1–61 (2019), arXiv:1308.2729.
19. Y. Rinott, T. Shoham, and G. Kalai, Statistical aspects of the quantum supremacy demonstration (2020), arXiv:2008.05177.

20. E. Pednault *et al.*, Leveraging secondary storage to simulate deep 54-qubit Sycamore circuits (2019), arXiv:1910.09534.

21. Y. Zhou, E. M. Stoudenmire, and X. Waintal, What limits the simulation of quantum computers? (2020), arXiv:2002.07730.

22. C. Huang *et al.*, Classical simulation of quantum supremacy circuits (2020), arXiv:2005.06787.

23. S. Irani (Moderator), Supremacy Panel, Hebrew University of Jerusalem, Dec. 2019. Participants: D. Aharonov, B. Barak, A. Bouland, G. Kalai, S. Aaronson, S. Boixo, and U. Vazirani. You Tube https://youtu.be/_Yb7uIGBynU.

24. F. Wilczek, Physics in 100 years, arXiv:1503.07735 (2015).

25. Y. Aharonov and D. Bohm, Time in the quantum theory and the uncertainty relation for time and energy, *Physical Review* **122**, 1649–1658 (1961).

26. Y. Atia and D. Aharonov, Fast-forwarding of Hamiltonians and exponentially precise measurements, *Nature Communications* **8**, 1572 (2017), arXiv:1610.09619.

27. J. Preskill, Sufficient condition on noise correlations for scalable quantum computing, *Quantum Information and Computing* **13**, 181–194 (2013).

28. L. Polterovich, Symplectic geometry of quantum noise, *Communications in Mathematical Physics* **327**, 481–519 (2014).

29. M. Ozawa, Uncertainty relations for joint measurements of noncommuting observables, *Physics Letters A* **320**, 367–374 (2004).

30. P. Busch, T. Heinonen, and P. Lahti, Noise and disturbance in quantum measurement, *Physics Letters A* **320**, 261–270 (2004).

31. C. W. J. Beenakker, Search for non-Abelian Majorana braiding statistics in superconductors, arXiv:1907.06497.

32. B. Terhal, The fragility of quantum information?, in *Theory and Practice of Natural Computing. TPNC 2012*, eds. A. H. Dediu, C. Martín-Vide, and B. Truthe, Lecture Notes in Computer Science, Vol. 7505 (Springer, Berlin, 2012), pp. 47–56, arXiv:1305.4004.

33. S. Wolfram, Undecidability and intractability in theoretical physics, *Physical Review Letters* **54**, 735–738 (1985).

34. I. Pitowsky, The physical Church thesis and physical computational complexity, *Iyuun, A Jerusalem Philosophical Quarterly* **39**, 81–99 (1990).

35. S. Aaronson and S. Gunn, On the classical hardness of spoofing linear cross-entropy benchmarking (2019), arXiv:1910.12085.

36. W. A. Borders *et al.*, Integer factorization using stochastic magnetic tunnel junctions, *Nature* **573**, 390–393 (2019).

37. G. Kalai, The quantum computer puzzle (expanded version) (2016), arXiv:1605.00992.

38. G. Kalai and G. Kindler, Concerns about recent claims of a huge quantum computational advantage via Gaussian boson sampling, preprint 2021.

39. J. Renema, Marginal probabilities in boson samplers with arbitrary input states (2020), arXiv:2012.14917.

40. P. Clifford and R. Clifford, The classical complexity of boson sampling (2017) arXiv:1706.01260.

41. A. D. King *et al.*, Scaling advantage over path-integral Monte Carlo in quantum simulation of geometrically frustrated magnets, *Nature Communications* **12**, 1113 (2021).

42. H. Zhang *et al.*, Quantized Majorana conductance, *Nature* **556**, 74–79 (2018).
43. H. Zhang *et al.*, Retraction Note: Quantized Majorana conductance, *Nature* (2021).
44. F. Pan and P. Zhang, Simulating the Sycamore quantum supremacy circuits (2021), arXiv:2103.03074.
45. N. Harvey *et al.*, *Journal of Algorithms* **59**, 53–78 (2006).
46. R. P. Feynman, Simulating physics with computers, *International Journal of Theoretical Physics* **21**, 467–488 (1982).
47. P. W. Shor, Polynomial-time algorithms for prime factorization and discrete logarithms on a quantum computer, *SIAM Review* **41**, 303–332 (1999). (Earlier version, *Proceedings of the 35th Annual Symposium on Foundations of Computer Science*, 1994.)
48. P. W. Shor, Scheme for reducing decoherence in quantum computer memory, *Physical Review A* **52**, 2493–2496 (1995).
49. A. M. Steane, Error-correcting codes in quantum theory, *Physical Review Letters* **77**, 793–797 (1996).

Chapter 7

Laws and Norms as Social Institutions[*]

Partha Dasgupta

University of Cambridge

1. Introduction

The Oxford dictionary defines "institution" as "an established law, custom, usage, practice, organisation, or other element in the political or social life of a people." We shall follow that lead but recast it to stress the role of institutions in economic life. By institutions we shall mean, very loosely, the arrangements that govern collective undertakings. Those arrangements include not only legal entities, like the modern firm, but also communitarian rotating savings associations such as the *iddir* in Ethiopia. They include the markets where we purchase goods and services, and the rural networks that have households in poor countries as members. They include the nuclear household in the West and the extended kinship system of claims and obligations in Africa. And they include that overarching entity called government everywhere.

Institutions are defined in part by the rules and authority structure that govern collective undertakings, but also in part by the relationships they have with outsiders. The rules on the factory floor (who is expected to do which task, who has authority over whom, and so on) matter not only to members of the firm, but also to others too. For example, rich countries have laws relating to working conditions in factories. Moreover, environmental regulations constrain what firms are able to do with their effluents. In every society, there are layers of rules of varied coverage.

---

[*] This essay has been taken from P. Dasgupta, *Economics: A Very Short Introduction* (Oxford University Press, 2007).

Some rules come under other rules, many have legal force, while others are, at best, tacit understandings.

The effectiveness of an institution depends on the rules governing it and on whether its members obey the rules. The codes of conduct in the civil service of every country include honesty, but governments differ enormously as to its practice. Social scientists have constructed indices of corruption among public officials. One such index is based on the perception private firms have acquired, on the basis of their experience, of the bribes people have had to pay officials in order to do business. The index — which is on a scale of 1 (highly corrupt) to 10 (highly clean) — is less than 3.5 for most poor countries (African countries and Eastern Europe are among the worst), and greater than 7 for most rich countries (Scandinavian countries are among the best). It used to be argued that bribery of public officials helps to raise national income because it lubricates economic transactions. It does so in a corrupt world: if you don't pay up, you don't get to do business. But corruption isn't an inevitable evil. There are several poor countries where corruption is low. Having to pay bribes raises production costs; so less is produced. Citizens suffer because the price they have to pay for products is that much higher.

Economists have speculated that government corruption is related to the delays people face in having the rule of law enforced. The thought is that delays are a way of eliciting bribes to hasten legal processes. In the early years of this century, it took 415 days in poor countries to enforce a contract, as against 280 days in the rich world. It may be that corruption is also related to government ineffectiveness. Registering a business takes 66 days in the poor world, 27 days in the rich world. In poor countries, registering a property takes 100 days on average, while in rich countries the figure is 50 days. Some economists have suggested that government officials in poor countries create lengthy queues (that is government ineffectiveness) to elicit bribes from applicants if they want to jump those queues (that's corruption).

How do government corruption, ineffectiveness, and indifference to the rule of law translate into the kind of macroeconomic statistics we have been studying here? They leave their imprint on an economy's "enabling capital". Other things being equal, a country whose government is corrupt or ineffective, or where the rule of law is not respected, is a country whose productivity of capital goods is lower than that of a country whose government suffers from fewer of those defects. Some scholars call these intangible but quantifiable factors "social infrastructure", others call them "social capital".

Institutions are overarching entities. People interact with one another in institutions. A more basic notion is that of engagements among people. The possibility of engagements gives rise to a fundamental problem in economic life: how to create and sustain trust among people and in their institutions. We study that next.

2. Trust

Imagine that a group of people have discovered a mutually advantageous course of actions. At the grandest level, it could be that citizens see the benefits of adopting a Constitution for their country. At a more local level, the undertaking could be to share the costs and benefits of maintaining a communal resource (irrigation system, grazing field, coastal fishery); construct a jointly usable asset (drainage channel in a watershed); collaborate in political activity (civic engagement, lobbying); do business when the purchase and delivery of goods cannot be synchronised (credit, insurance, wage labor); enter marriage; create a rotating saving and credit association (*iddir*); initiate a reciprocal arrangement (I am helping you, now that you are in need, with the understanding that you will help me when I am in need); adopt a convention (send one another Christmas cards); create a partnership to produce goods for the market; enter into an instantaneous transaction (purchase something across the counter); and so on. Then, there are mutually advantageous courses of action that involve being civil to one another. They range from such forms of civic behaviour as not disfiguring public spaces and obeying the law more generally, to respecting the rights of others.

Imagine, next, that the parties have agreed to share the benefits and costs in a certain way. Again, at the grandest level, the agreement could be a social contract among citizens to observe their Constitution. Or, it could be a tacit agreement to be civil to one another, such as respecting the rights of others to be heard, to get on with their lives, and so forth. Here, we will be thinking of agreements over transactions in goods and services. There would be situations where the agreement was based on a take-it-or-leave-it offer that one party made to another (as when someone accepts the terms and conditions set by a supermarket when making a purchase there). In other contexts, bargaining may have been involved (as when someone purchases household fineries at the rural fair in South Asia, which is not altogether different from a middle-eastern bazaar). The question arises: *Under what circumstances would the parties who have reached agreement trust one another to keep their word?*

Because one's word must be credible if it is to be believed, mere promises wouldn't be enough. (Witness that we warn others — and ourselves too — not to trust people "blindly".) If the parties are to trust one another to keep their promise, matters must be so arranged that: (**a**) at every stage of the agreed course of actions, it would be in the interest of each party to plan to keep his or her word if all others were to plan to keep their word; and (**b**) at every stage of the agreed course of actions, each party would believe that all others would keep their words. If the two conditions are met, a system of beliefs that the agreement will be kept would be self-confirming.

Notice that condition (**b**) on its own would not stand. Beliefs need to be justified. Condition (**a**) provides the justification. It offers the basis on which everyone could, in principle, believe that the agreement will be kept. A course of actions, one per party, satisfying condition (**a**) is called a "Nash equilibrium", in honour of the mathematician John Nash — he of *The Beautiful Mind* — who proved that it is not a vacuous concept. (Nash showed that the condition can be met in realistic situations.) However, the way we have stated condition (**a**) isn't due to Nash, but John Harsanyi, Thomas Schelling, and Reinhard Selten, the three social scientists who refined the concept of Nash equilibrium, so that it could be applied to situations where Nash's own formulation is not adequate.

Notice that condition (**a**) on its own would not stand either. It could be that it is in each one's interest to behave opportunistically if everyone believed that everyone else would behave opportunistically. In that case non-cooperation is also a Nash equilibrium, meaning that a set of mutual beliefs that the agreement will not be kept would also be self-confirming.

Stated somewhat informally, a Nash equilibrium is a course of actions (*strategy*, in economic parlance) per party, such that no party would have any reason to deviate from his or her course of actions if all other parties were to pursue their courses of actions. As a general rule, societies harbour more than one Nash equilibrium. Some yield desirable outcomes, others do not. *The fundamental problem every society face is to create institutions where conditions (a) and (b) apply to engagements that protect and promote its members' interests.*

Conditions (**a**) and (**b**), taken together, require an awful lot of coordination among the parties. In order to probe the question of which Nash equilibrium can be expected to be reached — if a Nash equilibrium is expected to be reached at all — economists study human behaviour that are *not* Nash equilibria. The idea is to model the way people form beliefs about how the world works, how people behave, and how they revise their beliefs on the basis of what they observe. The idea is to track the consequences of those patterns of belief formation to check whether the model moves toward a Nash equilibrium over time, or whether it moves about in some fashion or the other but not toward an equilibrium.

This research enterprise has yielded a general conclusion. Suppose the economic environment in a certain place harbours more than one Nash equilibrium. Which equilibrium should be expected to be approached — if the economy approaches an equilibrium at all — will depend on the beliefs that people held at some point in the past. It also depends on the way people have revised their beliefs based on observations since that past date. But this is another way of saying that history matters. The narrative style of empirical economics becomes necessary at this point. Model building, statistical tests on data relating to the models, and historical narratives must work together synergistically if we are to make progress in understanding our social world.

Mutual trust is the basis of cooperation. In view of what we have learnt about the multiplicity of Nash equilibria, we are now led to ask what kinds of institution are capable of supporting cooperation. To answer that, it will prove useful to classify the contexts in which the promises people make to one another are credible. There are four cases to consider, involving (1) mutual affection, (2) pro-social disposition,

(3) external enforcement, and (4) mutual enforcement. We consider them sequentially.

2.1. *Mutual affection*

Consider the situation where the people involved care about one another and it is commonly known that they care about one another. The household is the most obvious example of an institution based on affection. To break a promise that we have made to someone we care about, is to feel bad. So, we try not to do it. From time to time, though, even household members are tempted to misbehave. As people who live together can observe one another closely, the risk of being caught misbehaving is high. This restrains household members even when the temptation to misbehave is great.

That said, the household cannot engage in enterprises that require people of many and varied talents. So, households need to find ways to do business with others. The problem of trust reappears at the interhousehold level. This leads us to search for other contexts where people can trust one another to keep their word.

2.2. *Pro-social disposition*

One such situation is where people are trustworthy, or where they reciprocate if others have behaved well toward them. Evolutionary psychologists have suggested that we are adapted to have a general disposition to reciprocate. Development psychologists have found that pro-social disposition can be formed by communal living, role modelling, education, and receiving rewards and punishments (be it here or in the afterlife).

We don't have to choose between the two viewpoints; they are not mutually exclusive. Our capacity to have such feelings as shame, guilt, fear, affection, anger, elation, reciprocity, benevolence, jealousy, and our sense of fairness and justice have emerged under selection pressure. Culture helps to shape preferences, expectations, and our notion of what constitutes fairness. Those in turn influence behaviour, which are known to differ among societies. But cultural coordinates enable us to identify the situations *in* which shame, guilt, fear, affection, anger, elation, reciprocity, benevolence, and jealousy arise; they do not displace the centrality of those feelings in the human makeup. The thought we are exploring here is that, as adults, we not only have a disposition for such behaviour as paying our dues, helping others at some cost to ourselves, and returning a favour, we also ease our hurt by punishing people who have hurt us intentionally; and shun people who break agreements, frown on those who socialise with people who have broken agreements, and so on. By internalising norms of behaviour, a person enables the springs of his actions to include them. In short, he has a disposition to obey the norm, be it personal or social. When he does violate it, neither guilt nor shame would typically be absent, but frequently the act will have been rationalised by him. Making a promise is a commitment for that person; and it is essential for him that others recognise it to be so. The mutual influence of the sense of citizenship and being members of civil society may be why civics used to be taught at school. Rights and obligations do not always have to be enshrined in the law.

People are trustworthy to varying degrees. When we refrain from breaking the law, it is not always because of a fear of being caught. The problem is that although pro-social disposition is not foreign to human nature, no society could rely exclusively on it. How is one to tell "to what extent someone is trustworthy"? If the personal benefits from betraying one's conscience is large enough, almost all of us would betray it. Most people have a price, but it is hard to tell who comes at what price.

Societies everywhere have tried to establish institutions where people have the incentives to do business with one another. The incentives differ in their details, but they have one thing in common: *those who break agreements without cause are punished.* Below we study and see how that is achieved.

2.2.1. *Laws and norms*

There are two ways. One is to rely on an external enforcer, the other on mutual enforcement. Each gives rise to a particular type of institutions. Depending on the nature of the business they would like to enter into, people invoke one or the other. The coded term for one is the *rule of law*; for the other, it is *social norm*. People in the rich world rely heavily on the former, while in the poor world people depend greatly on the latter. Subsequently we will study the claim that it is *because* they have been able to depend extensively on the former for centuries that people in the rich world are now rich.

We illustrate the two methods of enforcement with the help of a numerical example of bilateral agreement. The numbers will allow us to draw insights without fuss. The example itself is based on the "putting-out system" of production, widely practised in Europe in the 17th and 18th centuries and is prevalent in poor countries today in the crafts. The system amounted to a patron-client relationship, but for our purposes here, it can also be thought of as a partnership.

Imagine that person A owns some working capital (raw material, say), worth $4,000 to him. A knows B, who has the skills to use that capital to produce goods worth $8,000 in the market. A doesn't have those skills. However, A has access to the market, which B doesn't. A proposes to advance his capital to her, with the understanding that he will sell the goods once B produces them and share the proceeds with her. If B were not to work for A, she would use her time to produce goods for her home, worth $2,000 to her. In order to get her to accept his offer, A proposes a sharing rule that is hallowed by their tradition: The $8,000 would be used first to compensate both parties fully — $4,000 for A (the amount A would enjoy from the best alternative use of his working capital, which economists call the working capital's *opportunity cost*) and $2,000 for B (which is the opportunity cost of B's time and effort); the remaining $2,000 would then be divided equally between the two. A and B would receive $5,000 and $3,000 respectively. Each would gain $1,000 from the arrangement.

B regards the proposal as fair, but is worried about one thing: Why should she trust *A* not to renege on the agreement by keeping the entire $8,000 for himself?

2.3. *External enforcement*

Here is one possible way to ensure that *B* could trust *A*: the agreement is enforced by an established structure of power and authority. In many societies, tribal chieftains, village or clan elders, and warlords enforce agreements and rules on disputes. Here we imagine that the external enforcer is the state and that the agreement is drawn up as a legal contract. We include on this list the implicit "social contract" among citizens not to break the law. However, if contracts are to offer a viable means of doing things, breaches must be *verifiable*; otherwise, the external enforcer would have nothing to go by if asked to rule on it. To be sure, lawyers make a handsome living precisely because verification is fraught with difficulties. Rough estimates suggest that in the US, expenditure on the legal profession (lawyers, judges, investigators), on people who work in insurance (loss adjusters, insurance agents), and on those in law enforcement (the police) make up $245 billion a year, which is about 2% of US's GDP; and we haven't included the defensive measures people take against possible litigations, burglary, and theft.

We leave aside the problems that arise in verifying breach of contract and note that if the punishment the state imposes for a violation is known to be severe, relative to the temptation *A* faces to violate, *A* will be deterred from going that route. If *B* is aware of the force of that deterrence, she will trust *A* not to renege. And *A* will trust *B* not to renege, because he knows *B* doesn't fear that he will renege. In the modern world the rules governing transactions in the marketplace are embodied in the law of contracts. A modern firm is a legal entity, as are the financial institutions through which employees are able to accumulate their retirement pension, save for their children's education, and so on. Employees have employment contracts with their firm. The agreements people reach with the saving and pension institutions are legal contracts. Even when someone goes to the grocery store, the purchases (paid in cash or by card) involve the law, which provides protection for both parties (the grocer, in case the cash is counterfeit or the card is void; the purchaser, in case the product turns out on inspection to be substandard). Formal markets, from which people enter and exist when they need to or wish to, can function only because there is an elaborate legal structure that enforces the agreements known as "purchases" and "sales". Moreover, it is because the customer, the grocery store's owner, and the credit card company are confident that the government has the ability and willingness to enforce contracts that they do business together.

Given that enforcing contracts involves resources, what is the basis of that confidence? After all, the contemporary world has shown that there are states and there are states. One answer — in a functioning democracy — is that the government worries about its reputation. A free and inquisitive press helps to sober the government into believing that incompetence or corruption would mean an end to its rule, come the next election. Notice how this involves a system of interlocking beliefs about one another's abilities and intentions. The millions of households in their country trust their government (more or less!) to enforce contracts, because they know that government leaders know that not to enforce contracts efficiently would mean being thrown out of office. In their turn, each side of a contract trusts the other not to renege (again, more or less!), because each knows that the other knows that the government can be trusted to enforce contracts. And so on. Trust is maintained by the threat of punishment (a fine, a jail term, dismissal, or whatever) for anyone who breaks a contract, be the contract legal (employment contract) or social (the contract between the voters and the government to maintain law and order). We are in the realm of beliefs that are held together by their own bootstraps (our earlier condition (**b**)).

What we have presented is only the sketch of an argument. The complete argument is similar to the one which shows that social norms also offer a way to enforce agreements. So, we turn to that fourth context in which agreements are kept; the play of social norms.

2.4. *Mutual enforcement*

Although the law of contracts exists today in every country, there are places where people can't depend on it. The nearest courts could be far away and there may be no lawyers in sight. When roads are decrepit, villages can be like enclaves. Much economic life is shaped outside a formal legal system. Nevertheless, people do business with one another. Saving for funerals in Ethiopia involves saying, "I accept the terms and conditions of the *iddir*." As there are no formal credit markets where they live, villagers practise reciprocity, so as to allow smooth consumption. A study found that in a sample of villages in Nigeria nearly all credit transactions were either between relatives or between households in the same village. No written contracts were involved, neither did the agreements specify the date of repayment or the amount repaid. Social codes were implicitly followed. Less than 10% of the loans were in default.

Why would the villagers in our example trust one another? They would, if agreements were mutually enforced: *a threat by members of a community that stiff sanctions will be imposed on anyone breaking an agreement would deter everyone from breaking it.* This is a common basis for doing business in the poor world. Among the Kofyar farmers in Nigeria, for example, agricultural land is privatised, but free-range grazing is permitted once the crops have been harvested. Kofyar households are engaged in subsistence farming; so labour isn't paid a wage. However, the Kofyars have instituted communal work on individual farms. Although some of this is organised in clubs of 8-10 individuals, there are also community-wide work parties. A household that doesn't provide the required quota of labour without good excuse is fined (as it happens, in jars of beer). If fines aren't paid, errant households are punished by being denied communal labour and subjected to social ostracism. In a different context, systems of codes have served to protect fisheries in coastal villages of northern Brazil. Violations are met with a range of sanctions that include both shunning and sabotaging fishing equipment. And so on.

How is mutual enforcement able to support agreements? It is all well and good to say that sanctions will be imposed on opportunists, but why should the threats be believed? They would be believed if sanctions were an aspect of social norms of behaviour. To see why, assume for the moment that whether an agreement has been kept by each party is *observable* by all parties. No doubt this is a strong assumption, but as with "verifiability", it is a useful starting point. Once we draw conclusions from it, we will be able to infer how communities could modify their institutions in situations where the assumption doesn't hold even approximately. That said, anyone who has visited villages in poor countries will know that there is less insistence on privacy there. In tropical villages, cottages are frequently designed and clustered in such a fashion that it must be hard for anyone to prevent others from observing what they are about.

By a social norm, we mean an accepted rule of behaviour. A rule of behaviour reads like: "I will do *X* if you do *Y*; I will do *Z* otherwise"; "I will do *P* if *Q* happens; I will do *R* otherwise;" and so forth. For a rule of behaviour to *be* a social norm, it must be in the interest of each person to act in accordance with the rule if all others act in accordance with it; that is, the rule should correspond to a Nash equilibrium. To see how social norms work, let us return to our numerical example to study whether cooperation based on a *long-term relationship* can be sustained between *A* (we now call him the patron) and *B* (we now call her the client).

Imagine that the opportunity for *A* and *B* to do business with each other is expected to arise over and over again; say, annually. The time taken for *B* to produce her output is assumed to be well within a year. Let *t* denote time. So, *t* assumes the values 0, 1, 2, ..., and so on, *ad infinitum*; with 0 standing for the current year, 1 standing for the following year, 2 standing for the year following that, and so on, *ad infinitum*. Although the future benefits from cooperation are important to both *A* and *B*, they will typically be less important than present benefits. After all, there is always the chance that one of the parties will not be around in the future to continue the relationship, or that circumstances may change in such ways that *A* does not have access to his capital flow. To formalise this idea, we introduce a positive number *r*, which measures the rate at which either party discounts the future benefits from cooperation. (We will see that in the present example it doesn't matter what *B*'s discount rate is. For expositional ease we assume that both individuals discount their future costs and benefits at the rate *r*.) The assumption is that, when making calculations in the current year (which is $t = 0$), each divides his or her benefits in any future year t by a factor $(1+r)^t$. The term $(1+r)^t$ denotes $(1+r)$ multiplied to itself t times.) So, if r is positive, $(1+r)^t$ exceeds unity for all future t; and since benefits in year t are divided by $(1+r)^t$ when making calculations in the current year, the importance of those benefits decays by a fixed percentage r each year, when viewed from today. The smaller is r, the greater is the weight placed on the benefits of future cooperation. We now show that, provided r is small, the pair could, in principle, enter a successful long-term relationship, where each year A advances \$4,000 to B, sells the goods that B has produced for \$8,000, and pays her \$3,000. The box below provides a formal argument.

The Grim Norm

Consider the following rule of behaviour that A might adopt: (a) to begin by advancing \$4,000 to B, (b) to sell the goods if she produces them during the year, (c) to share the proceeds according to the agreement, and (d) continue doing so every year, so long as neither party has broken the agreement; or (e) to end the relationship permanently the year following the first defection by either party. Similarly, consider the following rule of behaviour that B might adopt: so long as neither party has reneged on the agreement, worked faithfully for A each year; but refuse to ever work for A the year following the first violation of the agreement by either party.

The two rules embody a common idea: begin by cooperating and continue to cooperate, so long as neither party has broken their word, but withdraw cooperation permanently following the first defection from the agreement by either party. Withdrawal of cooperation is the sanction. Game theorists have christened this most unforgiving of rules the "grim strategy", or simply *grim*. We show next that grim is capable of supporting the long-term relationship, if r is not too large.

First consider *B*. Suppose *A* has adopted grim and *B* believes that he has. He will advance her the capital at the beginning of year 0. *B*'s best course of actions is clear: keep to the agreement. Suppose she reneges on the agreement. She would lose $1,000 (her share of $3,000 minus the $2,000 she would earn by producing home goods), but gain nothing in any future year (remember, *A* has adopted grim). This means that no matter what *B*'s discount rate is, she couldn't do better than to adopt grim, if *A* has adopted grim.

The harder piece of reasoning is *A*'s. Suppose *B* has adopted grim and *A* believes she has. If he has advanced the working capital to her, she will have worked faithfully for him in year 0. *A* now wonders what to do. If he reneges on the agreement, he would make a $4,000 profit ($8,000 minus the $4,000 he could have earned with his capital even if he had not entered into the relationship with *B*). But since he believes *B* to have adopted grim; he must also believe that *B* will retaliate by never working for him again. So, set against a single year's gain of $4,000 is a net loss of $1,000 (the foregone profit from the partnership) every year, starting year 1. That loss, calculated in year 0, is the sum, $(1,000/(1+r) + 1,000/(1+r)^2 + 1,000/(1+r)^3 + \ldots$ *ad infinitum*), which can be shown to add up to $1,000/r. If $1,000/r exceeds $4,000, it isn't in *A*'s interest to break the agreement, which means that he can't do better than to adopt grim himself. But $1,000/r exceeds $4,000 if and only if r is less than 1/4, or 25% (per year). We have therefore proved that if r is less than 25%, it is in each party's interest to adopt grim if the other party adopts grim. But if both adopt grim, neither would be the first to defect, which implies that the agreement would be kept. We have therefore proved that grim can serve as a social norm to maintain a long-term relationship between the patron (*A*) and the client (*B*).

Economists have found evidence of grim in social interchanges, but it would appear to be in force mostly where people also have access to formal markets. Mostly, grim is not in evidence. Sanctions are graduated, the first misdemeanour being met by a small punishment, subsequent ones by a stiffer punishment, persistent ones by a punishment that is stiffer still, and so forth. How are we to explain this?

Where formal markets and long-term relationships co-exist, grim could be expected to be in operation. Grim involves permanent sanctions, which is a needed device for preventing people from engaging in opportunistic behaviour when good, short-term opportunities appear nearby, from time to time. But if there are few alternatives to long-term relationships, communitarian arrangements would be of high value to all. Adopting grim would be an overkill in a world where people discount the future benefits from cooperation at a low rate. For that reason, the norms that are adopted involve less draconian sanctions than grim. A single misdemeanour is interpreted as an error on the part of the defector, or as "testing the water" (to check if others were watching). This is why graduated sanctions are frequently observed.

Here then is our general finding: *social norms of behaviour are able to sustain cooperation if people care sufficiently about the future benefits of cooperation.* The precise terms and conditions will be expected to vary across time and place; what is common to them all is that cooperation is mutually enforced, it isn't based on external enforcement.

3. Communities and Markets

How did people who now interact with one another get to connect in the first place? In traditional societies the answer is simple: mostly they have known one another from birth. People engaged in long-term relationships based on social norms — *communities*, for short — have to know one another, at least indirectly through people they know personally. Each person knows those with whom the local commons are shared. Communities are *personal* and *exclusive*. Members have names, personalities, and attributes. An outsider's word isn't as good as an insider's word.

In contrast, the hallmark of transactions enforced by the law of contracts is that they can take place among people who don't know one another. In the modern world people are mobile, a pattern of behaviour not unrelated to the fact that they are able to do business even with people they don't know. We all too often don't know the salespersons in the department stores where we are shopping, neither do they know us. When someone borrows from their bank, the funds made available to them come from unknown depositors. Literally millions of transactions take place each day among people who have never met and will never meet. Often, the exchanges take place only once, unlike exchanges based on long-term relationships. *Markets* are prime examples of institutions offering such opportunities. In contrast to communities, markets are *impersonal* and *inclusive*. Witness the oft-used phrase: "My money is as good as yours."

4. Trust as Self-Confirming Beliefs

There is, however, a piece of bad news: people could end up not cooperating even if they care a lot about the future benefits of cooperation. To see how, imagine that each party believes that all others will renege on the agreement. It would then be in each one's interest to renege at once, meaning that there would be no cooperation. Even if r is less than 25% in our numerical example, behaviour amounting to non-cooperation is also a Nash equilibrium: A doesn't advance the $4,000 worth of raw material to B, because he knows that B won't work for him; she would refuse because of the fear that A won't keep his promise to share the proceeds; a fear that is justified, given that A intends not to share the $8,000 with her once she has produced those goods; and so on. Failure to cooperate could be due simply to an unfortunate pair of self-confirming beliefs, nothing else. No doubt it is mutual suspicion that ruins their chance to cooperate, but the suspicions are internally self-consistent. In short, even when appropriate institutions are in place to enable people to cooperate, they may not do so. Whether they cooperate depends on mutual beliefs, nothing more.

Could the pair form a partnership if r exceeds 25%? The answer is "no". As grim is totally unforgiving, no other rule could inflict a heavier sanction for a single misdemeanour. The temptation A faces to defect is less if B adopts grim than if she were to adopt any other rule of behaviour; which implies that no rule of behaviour could support a partnership if r exceeds 25%. Studying grim is useful, because it allows us in many examples, such as the present one, to determine the largest value of r for which cooperation is possible.

We now have in hand, a tool to explain how a community can skid from cooperation to non-cooperation. Ecological stress — caused, for example, by increasing population and prolonged droughts — can result in people fighting over land and natural resources. Political instability — in the extreme, civil war — could in turn be a reason why both A and B become concerned that A's source of capital will be destroyed or confiscated. A would now discount the future benefits of cooperation with B at a higher rate. Similarly, if the two fear that their government is now more than ever bent on destroying communitarian institutions in order to strengthen its own authority, r would rise. For whatever reason, if r were to rise beyond 25%, the relationship would break down. Mathematicians call the points at which those switches occur, *bifurcations*. Sociologists call them *tipping points*. Social norms work only when people have reasons to value the future benefits of cooperation.

Contemporary examples illustrate this. Local institutions have been observed to deteriorate in the unsettled regions of sub-Saharan Africa. Communal management systems that once protected Sahelian forests from unsustainable use were destroyed by governments keen to establish their authority over rural people. But Sahelian officials had no expertise at forestry, nor did they have the resources to observe who took what from the forests. Many were corrupt. Rural communities were unable to switch from communal governance to governance based on the law: the former was destroyed, and the latter didn't really get going. The collective vacuum has had a terrible impact on people whose lives had been built round their forests and woodlands.

Ominously, there are subtler pathways by which societies can tip from a state of mutual trust to one of mutual distrust. Our model of the partnership between A and B has shown that when r is less than 25%, both cooperation and non-cooperation are equilibrium outcomes. The example therefore tells us that *a society could tip over from cooperation to non-cooperation owing merely to a change in beliefs*. The tipping may have nothing to do with any discernable change in circumstances; the entire shift in behaviour could be triggered in people's minds. The switch could occur quickly and unexpectedly, which is why it would be impossible to predict and why it would cause surprise and dismay. People who woke up in the morning as friends would discover at noon that they are at war with one another. Of course, in practice there are usually cues to be found. False rumours and propaganda create pathways by which people's beliefs can so alter that they tip a society where people trust one another to one where they don't.

The reverse can happen too, but it takes a lot longer. Rebuilding a community that was previously racked by civil strife involves building trust. Non-cooperation doesn't require as much coordination as cooperation does. Not to cooperate usually means to withdraw. To cooperate, people must not only trust one another to do so, they also have to coordinate on a social norm that everyone understands. That is why it's a lot easier to destroy a society than to build it.

How does an increase or decrease in cooperation translate into macroeconomic statistics? Our numerical example captured a salient point, that an increase in cooperation raises wealth by permitting a more efficient allocation of resources: A's working capital was put to better use under cooperation, as was B's labour. Consider now two communities that are identical in all respects, excepting that in one people have coordinated at an equilibrium where they trust one another, while people in the other have coordinated at an equilibrium where they don't trust one another. The difference between the two economies would be reflected in the economy's productivity figures, which would be higher in the community where people trust one another than in the one where they don't. Enjoying greater income, individuals in the former economy are able to put aside more of their income to accumulate capital assets, other things being equal. So, asset accumulation there is higher. Mutual trust would be interpreted from the statistics as a driver of economic growth.

Chapter 8

Complexity in Energy and a Low Carbon Transition

Martin Freer

School of Physics and Astronomy, University of Birmingham,
Birmingham, B15 2TT, UK
M.Freer@bham.ac.uk

This is a contribution that examines the difficulties in transitioning an energy system from a historic high carbon emissions version to a system that meets the need to be zero net carbon by 2050. The contention is that national-scale intervention through a market-driven approach is a weakly coupled approach, which is unpredictable and difficult to manage. It is possible that local energy solutions can accelerate the low carbon transition, but come with their own risks. In particular, the national system provides a level of security and resilience which is hard to reproduce at the local scale.

1. Introduction

The truth of humankind's impact on the global climate has become more or less the accepted truth, but remains inconvenient for nations and businesses. The Fifth Assessment Report (AR5) of the United Nations Intergovernmental Panel on Climate Change, IPCC was unequivocal in its assertion that global warming and sea-level rises were happening and that there is a clear human influence on the climate.[1] The subsequent *Special Report on Global Warming of 1.5°C* (2018) observed that "Global net human-caused emissions of carbon dioxide (CO_2) would need to fall by about 45 percent from 2010 levels by 2030, reaching 'net zero' around 2050" in order to limit global warming to within 1.5 degrees [2]. The cost of action is estimated to be ~$ 2.4 trillion, which is approximately 2.5% of world GDP.[2]

Beyond rises in the sea level, there is a potential impact on food production as climate variation impacts the location of the optimal zone for crop production, with rice being a particular case in point,[3] where there is a marked sensitivity to the temperature of the germination of rice. The impact of global temperature rises will then not just be in terms of displacement of people from regions close to sea level, but socioeconomic. A report from the U.S. government on the impacts of climate change on society indicates that unless action is taken, climatological events could grow to cost the country nearly half a trillion dollars annually.[4]

The warning signs of climate change and global warming go back at least to the 1970s,[5] yet real action has been slow even in the face of mounting evidence. The slow pace of intervention at the scale required is curious and is almost certainly due to a combination of factors. These would include the cost of intervention, creation of stranded assets, economic impact, and social inertia. However, the complexity of the energy system is a further dimension that inhibits central decision-making and creates a reliance on the market to deliver organic change. This would seem to be a significant issue within the UK energy system. The increasing interest from regions, cities, and communities to take energy matters into their control presents an interesting twist, which may liberate a more dramatic transition with local and more informed decision-making. Alternatively, the diversification that local solutions bring may create a patchwork of regional energy solutions which themselves are difficult to bind into a national energy system with challenges to resilience and sustainability.

The following is a UK-centric narrative, which focuses on the transition of the City of Birmingham which declared a Climate Emergency in 2019 with a desire to reach net zero carbon emissions by 2030. It examines some of the barriers to transition at both the national and regional levels.

2. The UK Energy System

Since the 1970s the UK's annual electricity demand has grown from ~250 TWh to above 350 TWh. This growth was originally provided by a mixture of coal, oil, and nuclear-fueled plants. The move by the British

Government to close UK coal mines in the 1980s was driven by the belief that coal was uneconomic to mine rather than any commitment to climate action. The oil crisis in the Middle East in the preceding decade had already weakened the reliance on oil imports and the opening up of the North Sea gas fields provided the solution. By the time of 2008 and the UK Climate Change Act[6] electricity by coal production was down to ~30% with the remainder being made up of gas nuclear and a small contribution from renewables. The plan henceforth as articulated by the Committee on Climate Change in their *2011 Renewable Energy Review* was "Together with ongoing investment in onshore wind, this would result in a 2030 renewable generation share of around 40% (185 TWh). Sector decarbonisation would then require a nuclear share of around 40% and a CCS share of 15%, along with up to 10% of generation from unabated gas."[7] (one of several potential scenarios).

Rather dramatic progress has been made in terms of wind power which sees over 20 GW of installed generation capacity (~50% of the average peak demand) with about 8.5 GW of offshore wind power. The pace of change within the decade since 2008 has been astonishing in this sector and is set to continue. Similarly, there has been a growth in solar power with over 3% of generation coming from this source. By comparison, the amount of installed generation from coal is now less than 9 GW and is set to fall to zero close to the middle of the 2020s, with already extended periods of coal-free generation. The instability of renewables has to date been compensated for by increasing the generation by gas with >30 GW of generation capacity.

The result of the above is that the UK energy system has become a great deal greener, though not zero carbon — which is now the UK government's target by 2050. It could read as a triumph for government-led energy policy, central planning, and the ability to steer the markets via incentives and mechanisms such as long-term contracts with a negotiated electricity price through the Contracts for Difference, CfD, approach. This may be the correct interpretation, but only time will tell and there are reasons to have doubts. The Committee on Climate Change[8] had suggested that a significant nuclear contribution was part of the mix, yet this has been incredibly hard to deliver. The high cost of nuclear construction, even though the payback is over the 60-year operation span of the plant, has been extremely challenging. The circles of negotiation and challenge associated with getting the EDF Hinkley Point C EPR reactor to the point of construction were dizzying. The risk for EDF/AREVA is enormous given the investment cost and the hurdles such as the State Aid enquiry of the European Commission should have been enough to deter future developments. Indeed, the Japanese Advanced Boiling Water reactor project has evaporated as has the potential development of the Westinghouse designed AP1000 reactor. The reliance of the UK government on market mechanisms and the CfD mechanism has been too weak to clear the barrier to confidence and investment. Similarly, plans for significant investment in carbon capture and storage related to large-scale plants have more or less been shelved and hence there is still a strong reliance on unabated gas generation. Thus, there are only two strands of the reconfiguring of the energy system that have been successful — an increase in wind power and a decrease in coal. The result is not zero carbon.

The desire for market-led solutions has not been entirely successful and the political need to be seen to be on the side of the consumer as far as energy prices are concerned and the real issue of fuel poverty has limited the amount of investment by business into new infrastructure and the pace of change. At present it is very hard to see how nuclear energy is going to come anywhere near 40% and how gas generation is going to be removed from the grid. The latter would require a significant investment in energy storage technologies and these are only making a slow journey to market with rather limited investment from the government into early-stage demonstration projects. At present there is little sign that the market will pull these through as the business model does not stack up. Frustration and differences in political ideology could lead to the renationalisation of energy infrastructure, such as the electricity and gas networks.

Much of the above misses the bigger picture, namely that electricity is only part of the energy system and that more fully it includes transportation and heat. The plans for transportation rely largely on electrification of cars and vans, with perhaps hydrogen fuel cells for trucks and trains. This will require additional electrical generation which has yet to be properly accounted for and a large infrastructure rollout. Heat is very challenging as a great deal of the load is domestic, with a housing stock that suffers from poor thermal insulation. At present solutions vary from district heating using hot water/steam-based systems, heat pumps using either gas or electricity, or perhaps replacing the methane in the gas grid with hydrogen. The latter does not seem to be an immediate possibility due to the need to generate carbon emissions-free hydrogen. The present solutions for heat look like placing an even stronger reliance on electricity generation. As such the future of the UK energy system involves a strong interdependency between electricity, heat, and transportation.

Though there is ground for optimism through the large shift in renewables, it is very hard to be confident that there is a market-led solution for the UK, despite the UK government over several cycles of Prime Minister being committed to the UK's leadership in this policy area. There is also a nervousness that market interventions such as that associated with solar power can be counterproductive. The complexity of the energy system, the unpredictability of the markets, and the very large challenge of the infrastructure rollout, especially with heat, present very real challenges to a centrally coordinated, but not controlled, system.

In the modern era, it is tempting to think that by throwing the problem at a system with a large enough computing system, equipped with genetic algorithms and artificial intelligence, an optimised solution would emerge. There is, however, a fundamental question as to how to check and understand the validity of the solution but more deeply there is the issue that the solution is only as good as the data provided and it is well known that the data are patchy and unreliable.

The pressing awareness of the Climate Emergency, frustration with the pace of change, and a desire to own the problem and solution have led to regional approaches to local energy. Across the UK, cities are examining the potential to develop local infrastructure and, through energy companies, deliver energy both to business and domestic users. This may unlock a pace and scale of change that has been frustrated nationally. Locally, it is easier to collate and manage data, which has the potential to facilitate local decision-making. With this new local activism comes opportunity and challenge.

3. Local Energy and the West Midlands

Many UK cities, such as Bristol, Leeds, Manchester, and Nottingham are actively exploring developing integrated energy, transport and indeed waste processing solutions. Birmingham has installed an extensive heat network linking many of the city council buildings, the New Street railway station, and the Aston University campus. It is procuring the installment of electric vehicle charging infrastructure and hydrogen buses and is developing a clean air zone. Remarkably, it is achieving this on a shoestring budget but has been assisted by a series of regional stakeholders, from academia to business, with a desire to deliver regional benefits in terms of the energy system, jobs, and economic stimulus.

In 2017, within the political framework of the West Midlands, the West Midlands Combined Authority (WMCA), the organisation called Energy Capital was established. Now fully integrated into the WMCA, this organisation is a regional stakeholder body charged with oversight, and where required, organisation of regional energy developments. One of its first activities was to establish a regional Policy Commission to examine how best to facilitate regional/local energy development. This Commission, led by Sir David King,[8] recommended the formation of a series of Energy Innovation Zones. The thinking being that by focusing innovation and development on a few local test sites, learning and best practice could then be transferred to other parts of the region. This is the ultimate reductionism, where there is a chance of understanding the full dimensionality of the problem, the data and complexity, and producing optimised and regionally accepted and acceptable solutions. The report recommended four sites (now extended to five) with very diverse characteristics. One is associated with the development of infrastructure related to a large automotive manufacturer, airport, and new housing development. Another revolves around the energy infrastructure of the City of Coventry. The third is dedicated to the manufacturing needs of the more traditional metalworking industry of the Black Country. Finally, one is associated with large-scale energy-from-waste plants embedded in the city of Birmingham.

Subsequently, each of the Energy Innovation Zones has developed under the umbrella of Energy Capital, with a fifth being introduced that focuses on the development of an old coal power station in the North of the region. This regional experiment is mid-course, and the jury is out as to whether it can deliver the stimulus for the regional energy transition. However, the case study below focuses on probably the most advanced of these: Tyseley Energy Park.

4. Tyseley Energy Park Energy Innovation Zone

Figure 1 shows a schematic of the Tyseley Energy Park Energy Innovation Zone. It is situated within 5 km of the centre of the City of Birmingham and is home to the city's energy-from-waste plant. This plant consumes 350,000 tonnes of waste annually and generates approximately 25 MW of electricity. This electricity is fed into the electricity grid. Adjacent to it is a biomass plant that converts wood waste through a gasification process, eventually producing 10 MW of electricity. Both plants are sources of waste heat which could potentially be coupled to the city's district heating system. The next stage of development is to generate transport fuels from electricity. Already the private wire connection from the biomass plant is used for fast charging of electric vehicles, and a hydrogen electrolyser is being installed to provide hydrogen for public bus routes. On site is the production of biofuels, e.g. biodiesel, from pyrolysis methods, and a range of other investments are being investigated, such as the extension to support hydrogen-powered rail for the city.

As such, the Energy Innovation Zone is linking the opportunities for integrating energy, transport, and waste. There is an opportunity for the integration of energy storage solutions, which are key for balancing generation and demand both on the local and national scale. The vehicle for the creation of the park is a co-creation group involving political, industrial, and academic stakeholders, where decision-making occurs through discussion and all agree to work in the best interests of the park, the city, and the region.

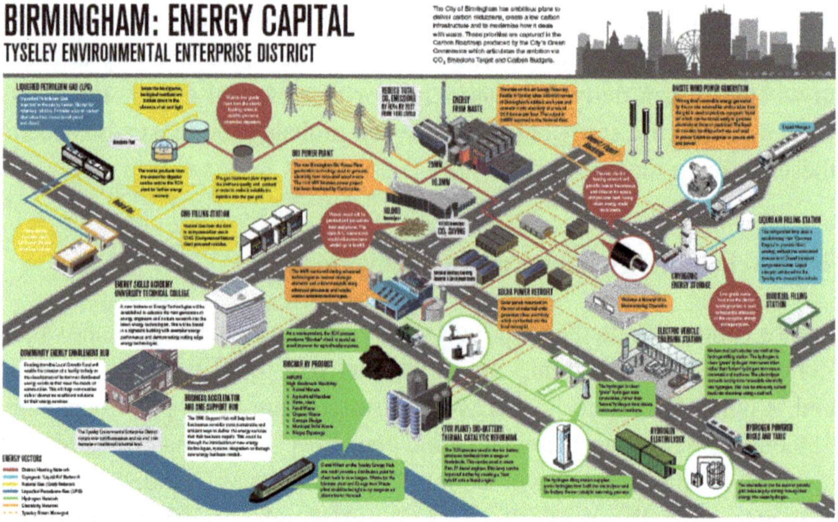

Fig. 1. The Tyseley Energy Park Energy Innovation Zone.

Although there is a measure of success to date, there are future obstacles to development. The park currently operates more or less as an island and to reach its full potential, it needs to be connected to the city's infrastructure. The first step here is the district heating network. However, the limited capacity within the city council, partitioning of responsibility for energy and waste, and the need to manage risk for the city council all pose challenges. The wider benefit to the local communities from the utilisation of heat and electricity will create further obstacles, as there is a need to navigate the UK's regulatory environment and ensure that domestic consumers do not suffer as a consequence. Nevertheless, this is the critical path for the Energy Innovation Zone concept to play a role in the regional energy transition.

5. Reflections

This contribution draws its motivation from the glacial progress (indeed, glaciers would appear to be melting faster) towards a shift in the global energy system, which would deliver decarbonisation on the 2050 timescale. Though the evidence for doing something is irresistible and the consequences of not doing anything are significant, both in financial and human terms, we are mostly still on the starting blocks. The UK has been enthusiastic advocate for intervention, being the first to put into law the Climate Change Act in 2008, and yet even the most enthusiastic country finds itself somewhere close to the beginning of the race.

The broader topic of the meeting touches on complexity, rules, and laws. The energy system is bound by all three. It is highly complex, multidimensional, patchy in terms of reliable data, and dynamic in space and time. Such is the complexity that it is hard to fully grasp, and one is left with high-level simplifications that oversimplify the detail. Though they provide a human-level understanding, it is difficult to conceive of a solution. The resort relies on computer models, which in turn are incomplete and simplifications. The temptation is to use the ubiquitous tool of AI to see what emerges, but again, to be reliable in optimisation, the detail must be available. The rules may be found in the rules of physics, energy transfer, and thermodynamics, which determine how components of the system can function optimally, and the laws are the overarching regulatory environment that places demands on generators and distributors, and rights for consumers. The optimisation has to also take into account the complexity of these rules and laws.

The energy markets have been extremely successful in delivering value to the customer through competition and optimisation of the market variables. The market may be adjusted and stimulated by incentives or penalties (e.g. carbon price), but until now it seems that more than a perturbation of the market is required for change. A quantum jump to a new location where the markets can again pursue their function of local optimisation is needed. The present refuge in a complex system is to rely on the markets to deliver change — there is a strong possibility this will not happen.

It is argued here that local energy solutions simplify the scale and dimensionality, take advantage of local knowledge, stakeholders, and community buy-in, and may thus provide a vehicle that removes some of the challenges described. An attractive proposition is a city digital twin, which is a live time representation of the energy and transport infrastructure that allows interventions to be tested before being implemented; the ultimate city planning and performance tool. However, for such a local approach to be successful, there is a need for regional capacity and expertise, an ability to flex regulation and laws, and above all, regional investment.

How to manage an energy infrastructure that recognises the grid scale system and does not leave stranded assets, properly integrates local solutions, and provides backup and resilience remains to be solved.

References

1. *The Fifth Assessment Report of the IPCC*,
    https://www.ipcc.ch/assessment-report/ar5/
2. *Special Report on Global Warming of 1.5 degrees C*, IPCC
    https://www.ipcc.ch/sr15/
3. K. Begcy, J. Sandhu and H. Walia, *Front. Plant Sci.* **9**, 1768 (2018).
4. Fourth National Climate Assessment, *Volume II: Impacts, Risks, and Adaptation in the United States*, https://nca2018.globalchange.gov/
5. W. Broecker, "Climatic Change: Are We on the Brink of a Pronounced Global Warming?", *Science* **189**, 460 (1975).
6. *The UK Climate Change Act 2008*,
    https://www.legislation.gov.uk/ukpga/2008/27/contents
7. Committee on Climate Change, *The Renewable Energy Review*,
    https://www.theccc.org.uk/publication/the-renewable-energy-review/
8. *Powering West Midlands Growth: A Regional Approach to Clean Energy Innovation*, https://www.energycapital.org.uk/powering-west-midlands-growth-regional-approach-clean-energy-innovation/

Chapter 9

PHYART@UoB:
Physics Meets Art at the University of Birmingham

William J. Chaplin

School of Physics and Astronomy, University of Birmingham, UK
w.j.chaplin@bham.ac.uk

We introduce and summarise the PHYART@UoB project[a], which is enabling and exploring active collaborations between artists from a variety of practices and scientists from the School of Physics and Astronomy at the University of Birmingham. Support from the Institute of Advanced Studies (IAS) at the University has been instrumental to the programme, and provides a model for how institutional support bridging diverse disciplines can enable and energise such work.

1. Introduction

Our project explores the collaborations taking place between artists, and scientists from the School of Physics and Astronomy at the University of Birmingham (UoB). The artists employ sound, film, sculpture, photography, gaming and dance to create new ways of thinking about and experiencing physics, as well as the relationship between science and art. The project is grounded in documentation and research, and is active in presenting work, showing impact and creating new educational and creative outputs that celebrate the relationship between art and science.

Key academics are Professor Bill Chaplin (project coordinator) [astrophysics; studies of stars and exoplanets], Professor Kostas Nikolopoulos [particle physics; studies of the fundamental nature of matter at CERN] and Professor Andreas Freise [astrophysics; gravitational waves].

---

[a] https://www.phyartuob.co.uk/

The importance of developing close, long-standing relationships — which includes regular "immersive" meetings, formal residencies (Leverhulme award), and co-supervision of one artist's PhD in Interdisciplinary and Collaborative Music Theatre — lies at the project's core, and has allowed us to chart the progress of how practice has been influenced over time. As we shall go on to discuss, the active support provided by the Institute of Advanced Studies (IAS) at the University has been instrumental to the programme, providing a working model for how such interdisciplinary work can be supported and encouraged.

2. Background and History

The project can trace its beginnings to the collaboration between sound artist Caroline Devine and Professor Bill Chaplin. Caroline is a sound artist and composer. Her work investigates the boundary between sound and music, exploring voices, signals and sounds that are ordinarily imperceptible or in some way absent. She has a particular interest in listening to the natural physical world and the use of space as a compositional parameter.

Caroline's compositional practice explores microtones, interference and the perception of sound (psychoacoustics), and she is interested in finding ways to interpret and communicate results from data both in electronic form — through use of tone generators — but also in a more physical way through immersive sound installations that open up new ways of listening. Sound provided the natural link to Bill's research in helioseismology and asteroseismology, the study of stars by observation of their natural oscillations.

The Sun, and other stars like it, make sound naturally in their interiors. These sound waves are trapped within the star. They reinforce, and the star resonates at a discrete set of harmonics, like a wind instrument. We cannot literally listen to the stars; but because stars are huge balls of hot gas, the trapped waves make the stars gently compress and relax in a rhythmic way: as a result, they oscillate in periodic manner.

Caroline was intrigued when she saw results from Bill's research group on oscillations of the Sun, a field we call *helioseismology* [e.g., see Ref. 1]. The spectrum of the Sun's resonances — made from observations of a network of automated telescopes called the Birmingham Solar-Oscillations Network (BiSON)[2] — looked all too familiar from her work on tones and resonances. She contacted Bill, not only in an effort to understand more, but also to explore the possibility of collaborating on together future work. It was from there that a longstanding collaboration began, one grounded not only in producing new art-science pieces but also in exploring new ways to communicate science and to explore the collaborative process when artists and scientist engage with one another.

Caroline began to incorporate helioseismology data from BiSON into a number of sound works, including *5 Minute Oscillations of the Sun* (2012), which was shortlisted for a BASCA British Composer Award in 2013, *Oscillate* for *Soundworks* at the Institute of Contemporary Arts (2012), and *Space Ham* for BBC Radio 3 (2013). They both also applied successfully to the Public Engagement Grant Scheme of the Institute of Physics for funds to support an art-science installation at the Birmingham Thinktank Science Museum. The installation, which had an extremely successful 6-month run, combined elements of Caroline's composition *Oscillate* with an animated display that explained the science underlying the composition.

From one collaboration, more followed in double-quick time. Our academics are at the time of writing working closely with seven artists and artistic companies.

3. Project Activities and Goals

Our project is producing new collaborative pieces and performances that explore the relationship between science and art. We are working together with the artists to create novel ways to reach new audiences, and present and discuss physics to the public. We have had impact on the artists, which has cascaded to impact on public audiences. Through our work, we endeavour:

- To use art to help people engage with science; and science to help enable engagement with art;
- To make people who are usually only interested in art think about and be aware of science, and vice versa, and to change views, perceptions

and attitudes of how art and science interact, for both audiences and practitioners; and

- To explore how the scientist influences the practice of the artist and vice versa, and how this can be demonstrated clearly to, and shape the views of, the audience, i.e., demonstrating that the interaction has a tangible influence on the enjoyment and appreciation of the piece.

Photographer and anthropologist Liz Hingley[b] has a key role in the project. She works primarily on long-term multi-platform projects that explore contemporary rituals, systems of belief and belonging around the world. She is an honorary research fellow at the University of Birmingham. Liz is working with the other artists to document, explore, and record the interactions, and to strengthen and further the collaborations and in so doing to raise awareness for the potential of creative exchanges between art and science. This is to enable both the artists and scientists to reflect on their work and methodologies in new ways, thus enriching practice.

3.1. *More on specific collaborations*

Here, we highlight collaborations with three of the artists where the Institute of Advanced Studies at the University has provided important support.

3.1.1. *Caroline Devine*

First, we continue the story of the collaboration with Caroline Devine. In 2014, Caroline won a *Leverhulme Artist in Residence* award to spend time as part of Bill's research group in Birmingham. The objective for the residency was to give Caroline the opportunity to immerse herself in the research, to understand better the underlying principles, and ultimately, to extend her practice beyond the Sun to make artworks incorporating oscillations data for stars other than the Sun, the field of *asteroseismology*.[3] The residency also led to the piece *Poetics of (Outer) Space* (2015)[c], presented in collaboration with the IKON Gallery, a

---

[b] http://lizhingley.com/

[c] https://www.thewire.co.uk/video/caroline-devine_s-poetics-of_outer_space/o=20

contemporary art venue in Birmingham, and as part of the 2015 University of Birmingham *Arts and Science Festival*. It used the resonances of several oscillating stars observed by the NASA *Kepler* Mission, which *Kepler* also discovered were hosting planets.

There is now a new international dimension to the collaboration, with the *Stellar Astrophysics Centre* (SAC) at Aarhus University, Denmark. To support this new work, Devine was appointed as the inaugural IAS Creative Fellow. A programme of activities arranged around this appointment will develop cross-institutional collaboration in the art-science domain. Caroline has visited Aarhus to discuss their role in the NASA *TESS* Mission, which Bill and his team are involved in too. TESS will form the focus of the next work with Caroline.

3.1.2. *Humanhood: Rudi Cole and Júlia Robert Parés*

Humanhood is a forward-thinking dance company run by co-artistic directors Rudi Cole and Júlia Robert Parés. They create bold choreographic pieces, working with musical and lighting directors, and taking inspiration for their work from a variety of disciplines. Rudi and Júlia have a particular interest in integrating concepts from physics and astronomy into their practice, and we have been working with them since 2017.

We have run several immersive interaction sessions[d] (see Fig. 1), where Bill and his colleagues have engaged in workshops at the dance studio, exchanging ideas both on the collaboration itself but also how we each work to explore the similarities and differences in how artists and scientists practice in their field. Most recently, the interactions have focused on the development of a new work called *Torus*, which Humanhood recently premiered and are now touring (nationally and internationally).

---

[d] https://vimeo.com/272336877

Fig. 1: Photos from one of the immersive sessions with Humanhood, December 2018, at the *DanceXchange* in Birmingham.

Bill has also been involved in several post-show talks with Rudi and Júlia. We have explored with the audience how artists and scientists can influence each other's practice, which goes to the heart of the collaboration, i.e., how enjoyment and the impact of the piece is influenced by the knowledge of the art-science interaction. Reactions at post-show talks have included:

"The collaboration between dance and science was very creative and enjoyable"
"Creative outside the box thinking"
"Love collaboration to express the love of science in movement…forward and positive"

On developing the collaboration, we have received plenty of encouragement, including:

"Go deeper and deeper: long-term brings unexpected results or nuances that spill into the work of both artists and scientists if truly open to the process"

3.1.3. *Juliet Robson*

Everything in the universe vibrates, from the smallest molecule to distant stars and yet the vast majority of frequencies remain hidden from our senses. Conceptual artist Juliet Robson[e] has been working with Bill on the overarching project *hertz* to produce an interactive and novel artwork that allows an audience to hear, see and engage with the patterns of resonating stars. Having first raised the pitch of the measured stellar frequencies — to bring them into the audible range — Juliet has used the resulting tones to drive the resonances of a metal plate on which coloured sand has been sprinkled. As the pitch is changed, so the sand shifts according to the plate's resonant vibrations into beautiful ordered patterns. These Star Machines, built with support from the University, are interactive; participants generate the patterns of real, observed stars, sprinkling the coloured sand and changing the driving tones.

---

[e] https://julietrobson.com/

Fig. 2: The Star Machines.

Fig. 3: Young attendees at the 2018 Oxford IF Science Festival enjoying the Star Machines.

We have run two successful interactive events with the Star Machines: one at the Oxford IF Science Festival in October 2018[f] (see also Figs. 2 and 3), and more recently (with support from the IAS), at the University of Birmingham *Green Heart Festival* in June 2019.

---

[f] https://youtu.be/748CLugS1bw

Some reactions have included:

"Some incredible vibrations — I'll be able to go home and tell the family I heard the stars sing in Newbury!"

"Mind-blowing Stuff — Resonating sounds on a different spectrum — just trying to imagine an auditory dictionary/directory of different stars — Could you potentially identify stars from sound alone? So many Questions!"

"*hertz* was really exciting! As an artist I really enjoyed being able to see the patterns created and how I might be able to work with those."

3.2. *Workshops and other activities*

In addition to working with each of the artists, we have initiated a variety of wider activities to explore the overarching theme of collaboration between scientists and artists. The Institute of Advanced Studies (IAS) at Birmingham has been instrumental in realising these activities.

We organised an interdisciplinary arts-science workshop hosted by the IAS in Birmingham in September 2015. Professor Geoffrey Crossick, University of London and Director Arts and Humanities Research Council's (AHRC) *Cultural Value Project*, and Professor Barry Smith, Director Institute of Philosophy, University of London and Leadership Fellow AHRC *Science in Culture Theme*, acted as the co-conveners with Bill. The event explored the factors and processes that lead to and allow interdisciplinary collaborations to initiate and develop productively. It involved participants from across the UK, with a variety of stakeholders, including artists and scientists, funders, commissioners and curators.

Bill, Caroline, and one of the other workshop participants, Jacob van der Beugel (a ceramic artist) also presented at the University-Based Institutes for Advanced Studies (UBIAS) network conference *UBIAS into impact: networking our academics to meet global challenges* held in June 2016, which involved leaders from University Institutes around the world.

In 2017, the IAS Workshop *Humanhood: Dance Meets Science* focused on the creative opportunities and challenges that arise in arts-science collaborations: again, how science can inform art and the arguably more tricky direction of how art can inform science. Our collaboration with Humanhood acted as the lens through which to explore these issues.

Humanhood shared choreographic sequences they have been developing during the early stages of the collaborative process, performed by the company dancers, giving an opportunity to witness the close collaboration and to see how it translates to movement with a group of dancers. The event provided insights into how we have communicated and collaborated in a way that has informed both our thinking and process, and involved significant attendee participation with active discussion and engagement.

Feedback and reflections from participants included:

"My own research is on the interface between performance and medicine, so there are/were resonances for me."

"I work on the body and am very interested in exploring how my work might aligned with the art form that uses the body as communication (dance)."

"I feel that artists have to move beyond the 'I have been inspired by…and here is my response,' and if possible move into more rigorous and specific points of contrast and comparison."

"What would it mean to use dance in teaching methods in physics and astronomy, for instance? What about dance in the physics lab? I am interested in the space that can be opened up between the disciplines, and what that exchange might look like or do?"

References

1. W. J. Chaplin, *Music of the Sun: The Story of Helioseismology* (Oneworld Publications, 2006).
2. S. J. Hale, W. J. Chaplin, G. R. Davies *et al.*, "Performance of the Birmingham Solar-Oscillations Network (BiSON)", *Solar Physics* **291**, 1 (2016).
3. S. Basu and W. J. Chaplin, *Asteroseismic Data Analysis: Foundations and Techniques*, Princeton Series in Modern Observational Astronomy (Princeton University Press, 2017).

Chapter 10

Laws about Laws

Alastair Wilson

School of Philosophy, Religion and History of Science,
University of Leeds, Woodhouse Lane, Leeds LS2 9JT, UK

School of Philosophical, Historical and International Studies,
Monash University, VIC 3800, Australia
a.j.j.wilson@leeds.ac.uk

Laws of nature have two characteristic features. They are general, in that they apply across all situations of a given kind — although they are typically restricted to particular domains. They are also modal, in that they apply across possible situations as well as actual situations. This simple account captures the core features of laws and their differences across distinct fields, and it helps to explain why laws are less prominent in some fields than in others. The most fundamental laws of physics are a special case, in that they are maximally general: they apply to all possible situations whatsoever. This provides a principled basis for a reductionist — or, to use a softer term, physicalist — view of nature. Any plausible reductionism, however, still recognizes a rich world of explanations beyond physics. In domains such as biology where laws retain important explanatory power, as well as in more anarchic domains such as history, physics is not and can never tell the whole story — and physics itself is part of the explanation for why that is so.

1. A Philosophical View of Laws of Nature

One major theme of this volume is the diversity of laws of nature across different scientific disciplines. In this chapter I aim to draw attention to some underlying features which unite laws from all disciplines, diverse though they are. I will argue that lawhood in the sciences can be captured within a simple framework, where laws are characterized as *modalized generalizations*. This framework leaves to the sciences themselves all the substantive questions about what sorts of laws — if any — hold sway within a given domain. It can help us to see that laws of different domains are complementary, to see how some domains have few or no laws, and to see why not all scientific explanations reduce to those of fundamental physics.

The framework I will describe does assign a privileged position to fundamental physics: the laws of fundamental physics are the most general laws (in that they apply to the widest range of possible situations), and they are the laws which hold no matter what (in that no possible situation violates them). But that does not mean that these laws are the source of all of our explanations. Quite the contrary: it has come to be widely recognized in the recent philosophy of science literature that non-fundamental, or 'high-level' laws are indispensable to the large majority of our scientific explanations and that no amount of fundamental physics, however sophisticated, can ever replace them. Elsewhere in our explanatory projects we may find no reliable laws at all, and then we regulate our inquiry in a much more patchwork way via models, heuristics, direct inference from bulk data, testimony and many other factors.

If such different laws are involved in such different ways in the practices of the various sciences, why do we have a single concept of laws at all? Is it merely a cultural accident or linguistic coincidence that the term 'law' recurs across the sciences, and in scientists' reflections on their own achievements? In addition to the chapters in the present volume (in which there is a clear selection effect in favor of reflection on laws and lawhood!) many of the most thoughtful scientists writing on the nature of their own field have cast the matter in terms of laws or principles. My own disciplinary specialism is philosophy of physics, in which the concept of lawhood has played an absolutely central role historically. Newton (1687) formulated his theory of mechanics in terms of laws of motion and referenced those laws in the title his masterwork *Philosophiae Naturalis Principia Mathematica*; the second law of thermodynamics was said by Arthur Eddington to hold "the supreme position among the laws of nature" (Eddington 1928); and Einstein was famously guided in formulating special and general relativity by his principles of relativity and of equivalence. Einstein also made a philosophical distinction between principle theories (such as special relativity) and constructive theories (Einstein 1919) — with laws playing central but distinct roles in each type of theory. Laws are also emphasized in more recent popular writings by physicists — think for example of Feynman in the title of his *The Character of Physical Law* (Feynman 1965), Weinberg in the subtitle of his *Dreams of a Final Theory: The Scientist's Search for the Ultimate Laws of Nature* (Weinberg 1992). But laws and principles have played equally central roles in the special sciences — Mendel's laws of genetics are perhaps the most familiar example. This central role played by lawhood across the different sciences is in my view no accident; in the next section I outline an account of laws which helps to explain their ubiquity in science.

2. Laws are Universal Modalized Generalizations

What is a law of nature? Before offering my positive account, I want to note some key criteria it ought to meet. The first is broad extensional adequacy: an account of laws should capture, by and large, the existing use of the term 'law' (and of related terms such as 'principle') by scientists in their descriptions of their own activity. The second requirement concerns the relevance of laws: an account of laws should help to explain why it is that we are so interested in laws, and why so much effort is expended to discover them. The third requirement concerns our knowledge of laws: laws ought to be the kind of thing which we can discover (or at least, obtain solid confirmation for) using the kinds of evidence which is in fact available to us. While they sound modest, these criteria have proven surprisingly difficult for an account of laws in science to collectively satisfy (van Fraassen (1989) argues that they in fact cannot be collectively satisfied, on the basis that requirements closely related to relevance and knowability are in tension). I think that the account I offer satisfies all the criteria, although it does so in part by offloading some of the harder questions.

As I understand them, laws of nature are statements about reality which have two characteristic features. Laws are *general* in that they apply across many situations, though not necessarily universally; their generalizing power can be restricted to particular domains. This generality is what enables science to be more than just an endless list of happenings, of one staccato fact after another. Laws are also *modal*, in that they generalize across possible situations as well as actual situations. They do not just say how all actual situations of a certain type happen to turn out, but how all possible situations of that type in fact turn out. This modal character is the source of the relevance of laws in guiding our beliefs about the world and our interventions upon it. Combining these two features, I will talk of laws as *modalized generalizations*, in the sense that they are generalizations extended to range over possible situations as well as actual situations.

Here is the basic schema for laws as modalized generalizations:

Basic schema: all possible Fs are Gs.

The schema is deliberately abstract and topic-neutral, enabling it to be applied across fields as diverse as fundamental physics (where the Fs might be systems of interacting particles or fields) and psychiatry (where the Fs might be patients presenting with some distinctive symptom). F and G may indeed refer here to any property whatsoever, including to probabilistic properties of having some objective probability of having some further feature (such as *having a 50% chance of producing a pea plant with purple flowers*).

The kind of possibility at work here is what I call *natural possibility*, but what is intended is the same thing that is often called *physical possibility* or *nomic possibility* by philosophers. Whereas natural possibility is often defined as what is compatible with the laws of nature, for avoidance of circularity I will reject this definition and instead take the notion of natural possibility as a basic one. For those interested in the underlying metaphysics, I recommend my book *The Nature of Contingency* (Wilson 2020), which links possibility itself to quantum theory; but no particular account of the nature of possibility will be presupposed here. What matters, for understanding the account I will offer, is that the reader grasps a distinction between things that really can happen as part of the course of nature with things that really can't. Ducks really can flap their wings; ducks really can't eat their own shadows. Unfortunately, not every case of natural possibility will be so easy to adjudicate; that's where science comes in.

In this chapter I will be focusing on scientific laws, so I will not consider alternatives to natural possibility in the analysis of lawhood. But there is substantial potential mileage from using alternative notions of possibility in the analysis of lawhood. Logical possibility is one example; here the laws will end up closely aligned with the axioms of the logical system in question. A case of special interest is a move to permissibility according to some human legal or social code; there seem to be good prospects for extending the account so that the laws of cricket, say, are interpreted as capturing what all possible correctly-administered games of cricket have in common, or the laws of etiquette are interpreted as capturing what all possible socially respectable behavior has in common. That would enable us to understand 'law' in criminal-justice and sporting institutions on the same broad model as scientific laws; the alternative is to understand the normative uses of 'law' as having a common conceptual root but having diverged in content from the scientific usage. (The etymology indicates that physical laws were so called by analogy with human justice, rather than vice versa.) Back, then, to the focus on scientific laws.

Applied unrestrictedly, the basic schema above might characterize the true underlying laws of a future fundamental physics. We don't have that theory, we may never have that theory, but most importantly even if we do discover that ultimate physical theory then we cannot simply expect to shake it and have the remaining laws of sciences other than physics fall out. Higher-level science would not be rendered obsolete even by a complete fundamental physics, for reasons I explore in Sec. 5. Accordingly, an adequate framework needs to allow for laws with restricted domains. In particular, generalizations which play central roles in higher-level sciences such as biology will be restricted to apply to only a subset of all the things there are. For example, pretty much every law in zoology will be restricted to apply to animals; a carnivorous plant or a robot vacuum cleaner are not within its intended scope.

We can make explicit the restricted element of the proposed account:

Restricted schema: All possible Fs within domain D are Gs.

Domains may be restricted in many different kinds of ways. The restriction might be to things of a certain kind — to molecules, to thoughts, to economies — and/or to things in a certain location — Singaporean, terrestrial, cosmic. The former kind of restriction would give rise to laws of higher-level sciences (molecular chemistry, psychology, and economics respectively); the latter would give rise to laws which apply only in particular places and times (characteristic laws of Singapore's microclimate, or the law that objects dropped on Earth accelerate downwards at 9.81 ms^{-2}, or to regions of the universe where the prevailing values of physical parameters resemble those of our own region).

Unless we place some limits on what restrictions give rise to laws, they will proliferate — there will be laws of Welsh mountains, laws of toy drums, laws of my breakfast. We might attempt to narrow down the range of restrictions which give rise to laws by requiring that they be natural in the sense of Lewis (1983), or by ruling out any reference to particular individuals, or by imposing some other requirement. Alternatively, we might opt for a full pluralism in which any modalized generalization is a law, and explain away our lack of interest in most laws in terms of our lack of interest in the relevant restriction; Woodward (2018) pursues a similar line. I cannot pursue this metaphysical question further here; fortunately for present purposes I don't need to. Instead, I will stay neutral by saying that all modalized generalizations are at least *candidate* laws of nature.

The simple account of laws that I have described in this section is flexible enough to fit the form of laws as they appear across physics, biology, and other scientific disciplines. In fact, I would conjecture that it is a fully general view of laws of nature:

Unity Conjecture:
All candidate laws of nature fit the modalized generalization account.

Variety Conjecture:
Anything fitting the modalized generalization account is a candidate law of nature.

In subsequent sections I will look in more detail at what features distinguish laws of different disciplines; in the process we will accrue some incremental support for these two conjectures.

3. Fundamental Laws of Physics

The most fundamental laws of physics are a special case, in that they are maximally general, ranging over all the natural possibilities whatsoever. They constrain all the other laws of the other sciences in that nothing can contradict a fundamental law of physics; if it's physically impossible, then it cannot be chemically or biologically or psychologically possible either. Everything that happens is within the scope of the fundamental laws of physics, and nothing can violate them. These facts about the fundamental laws provide a principled basis for a reductionist — or, to use a softer term, physicalist — view of nature.

The reductionist view of science has been stoutly defended by David Gross in his contributions to ICA 3 in Singapore and Birmingham, and in his chapter in this volume. But Gross takes his reductionism much further than I think is plausible. In this section I would like to present an alternative form of reductionism, one which is less imperialistic in its implications for sciences other than physics. My more moderate reductionism still assigns a central metaphysical role to the laws of fundamental physics, but these laws are not taken to exhaust the explanations which science offers. The vast majority of the explanations science discovers will continue to remain backed by higher-level laws rather than by fundamental laws, and no amount of progress in fundamental physics will ever change that.

Fundamental laws of physics can be obtained from the Restricted Schema of Sec. 2 by imposing no restriction whatsoever; therefore the Basic Schema would in fact be adequate to fundamental laws of physics, if those were all we were interested in capturing. However, to do that would be to miss the underlying commonality between fundamental laws of physics and the laws of the non-fundamental sciences, and this commonality is revealed by the Restricted Principle. The restriction clause is vacuously satisfied in this case.

A contrast between laws of physics and principles of physics in a broader sense is sometimes drawn, for example by Richard Feynman and more recently by Marc Lange. Here is Feynman:

> "[T]here are a large number of complicated and detailed laws, laws of gravitation, of electricity and magnetism, nuclear interactions, and so on, but across the variety of these laws there sweep great general principles which all the laws seem to follow". (Feynman 1967, p.59, quoted in Lange 2012, p.154.)

Feynman goes on to mention conservation principles as an example of the kind of principle which transcends individual laws of physics, even the fundamental ones: the conservation of energy would be true, on this line of thought, whatever fundamental forces were at work in the world. This distinction between laws and principles may be readily captured within the context of the modalized generalization theory of laws by varying the sense of possibility involved. To characterize principles, we appeal not to universal generalizations over naturally possible happenings (as we did for laws), but rather to universal generalizations over happenings which may be naturally impossible but which still respect certain general features of natural possibility. We might call these more exotic yet still well-behaved possibilities *extended natural possibilities*; such possibilities might include different force laws but still respect the principle of conservation of energy. This potential for application to the distinction between laws and principles highlights the flexibility of the modalized generalization view of laws.

Truly fundamental laws of physics may be sparser, and more flexible, than one might initially imagine. As David Gross's chapter explains, the Standard Model of particle physics is immensely predictively and explanatorily powerful, but it contains a number of parameters whose value is not given any theoretical explanation — including the masses of a number of particles including six quarks and three leptons and the Higgs, a number of 'mixing angles' and 'gauge couplings' determining the nature and strengths of the interactions between particles of different types, and parameters characterizing the vacuum state and mirror symmetry violation. It is extraordinarily impressive that the outcome of all known measurements can be explained using only those 19 parameters. But it is nevertheless puzzling that the parameters should take exactly the values they do and not any other value, when historically the phenomena studied by fundamental physics of previous eras (and many of the parameters characterizing those phenomena) have later turned out to be relatively neat and explicable in terms of the successor concepts employed by subsequent physics. This sense of puzzlement intensifies in the face of the apparent extreme 'fine-tuning' of a subset of these parameters. What makes these parameters fine-tuned is that a) there is no explanation for their value at a broader theoretical level and b) the slightest variation in them would have led to a universe very different from ours and wholly uninhabitable.

The physicists I have consulted tend to regard the free parameters in the Standard Model as currently unexplained, but without there being an in-principle barrier to their being explained in the further passage of physics. There is nothing about these parameter values in particular which prevents us from attempting to explain them in independent terms, by developing new physics beyond the Standard Model. Moreover, the fact of fine-tuning does give us prima facie (and limited, and defeasible...) reason to suspect that there might be explanations of the values of these parameters out there to be given. John Leslie uses the vivid example of the firing squad to make this point (Leslie 1989) — one ought to be surprised if one survives a firing-squad and suspect something like a conspiracy to save your life, *even though* the shooters all missing is not completely impossible and *even though* you would not have been around to be surprised if they had not missed. In the current context, the analogy of a conspiracy would be a mechanism to ensure that parameter values suitable for intelligent life arise. There are a number of proposed mechanisms of this kind, many involving some kind of multiverse, and here is not the place to dwell on the details. For a variety of different types of response to fine-tuning arguments, see the essays in Carr and Ellis (2007).

The important point for present purposes is that if there are after all explanations to be given of the values of the fine-tuned parameters — whether or not these explanations involve multiverses — then the Standard Model and its parameters will no longer be properly regarded as a fundamental theory: rather, the fundamental theory will consist in whatever equations fix the values of the underlying parameters, and give rise to the Standard Model as an emergent theory which is true around here but which is not true in all of the natural possibilities. Only the underlying fundamental theory holds across all of the natural possibilities whatsoever, and the prospect remains open that this fundamental theory will not have any free variables whose values must be fixed by experiment; some approaches in string theory do seem to have this feature. But at this point we must draw back from these speculations about future string theory to reiterate the main points of this section.

Firstly, the fundamental laws of physics are metaphysically privileged. They cannot be broken; they constrain all the other laws; they apply unrestrictedly across all of time and space (and beyond time and space, according to some of the more radical approaches to quantum gravity). They are the source of some of our most impressive and accurate explanations and predictions — the remarkably accurate prediction of the magnetic moment of the electron, for example, appeals to various features of our current best guess at the fundamental laws. But the explanatory power of the laws of fundamental physics is nevertheless limited and does not even extend to some of the most basic knowledge we have in fields like genetics, economics and psychology — as I argue in the next section.

4. Emergent Laws

I have expressed sympathy for a reductionism which gives the laws of fundamental physics the special metaphysical status of constraining all other laws and phenomena. Nothing can possibly happen which violates the fundamental laws of physics, and those laws apply always and everywhere; by contrast, all other laws (including non-fundamental laws of physics) apply only within some restricted domain. My reductionism, however, is moderate in that it acknowledges a rich world of explanations beyond the reach of physics. Fundamental physics is of course an important part of the story of reality — metaphysically speaking, the most important part — but the laws of fundamental physics cannot completely supplant the laws of the rest of the sciences, and the explanatory methods of fundamental physics cannot generalize to science as a whole.

It is a striking fact about science that the different sciences deal with characteristically different scales of size. Fundamental physics deals with both the very small and the very large; biology deals with intermediate-sized things like bacteria and mice, and meteorology deals with large things like clouds and atmospheres. Indeed, it's tempting to think that some sciences study the building-blocks of the things other sciences study, and this idea motivated the classical philosophy-of-science picture of levels of reality defended by Oppenheim and Putnam (1957). On their view, the most fundamental theories describe the smallest things, and less fundamental theories describe larger things which are composed out of the smaller things. One thereby reduces the higher-level theory to the lower-level theory by showing that the higher-level theory can be recovered from the application of the lower-level theory to the smaller things. This rather crude account of levels left much to be desired, and a subsequent six decades of philosophical work on the topic has produced a much more nuanced account of levels, dropping many of the assumptions about parts and wholes made by Oppenheim and Putnam.

A classic contemporary account of levels is given by Christian List (List 2019) who distinguishes multiple types of level system at work in the sciences: levels of description, levels of ontology, and levels of explanation. It is the last of these, levels of explanation, which is the focus of my case for the irreducibility of high-level explanations. Hard versions of reductionism fail not because biological systems are made of some non-physical stuff in addition to ordinary matter, but because the explanations of evolutionary biology cannot be reduced to the explanations of fundamental physics. That animals are made of atoms doesn't settle the explanatory question.

If there are different levels of scientific explanation, then (given some conceptual connections between lawhood and explanation which are widely accepted within the philosophy of science) where we expect to find laws of these different levels which back, or *mediate*, the explanations operative at that level. And that is exactly what we do find. In biology, a sprawling and multifaceted discipline, a multifaceted range of laws are employed: apart from the Mendelian laws of inheritance already mentioned, numerous laws have been proposed of variable mathematical precision and variable intended scope. Green and

Wolkenhauer (2013) distinguish different types of biological laws, operative at various sub-levels within biology: higher-order laws, optimality principles, design principles (including evolutionary design principles), and organizing principles.

One heuristic for understanding higher-level laws (frequently employed by philosophers although rarely in a fully self-aware fashion) is to simply pretend that the level you are interested in is fundamental, and ask what the fundamental laws would be for a possible world like that. What would be the laws governing water if it was a homogenous continuous fluid instead of a mixture of discrete molecular constituents? But this heuristic hinders more than it helps; a more realistic picture of the laws of a level emerges if we can instead think of higher-level laws as modalized generalizations concerning some more or less *abstract* subject-matters. Here the sense of 'abstract' involved is linked to the conceptual process of abstraction, where irrelevant specific details of particular cases are ignored to focus on some more general commonality between cases. Accordingly, the way to understand the higher-level laws of biology and the other special sciences is that the detailed underlying physics is being strategically set aside, rather than being imagined away.

Abstract explanations in the higher-level sciences can take a variety of forms. Many are causal explanations (perhaps even all are — this is a topic of much dispute within philosophy of explanation) where the causes are described at any level of description above that of fundamental physics. To say that a substance is a certain kind of chemical compound is to put some constraints on its underlying fundamental physical composition, but there are an enormous number of different fundamental physical states of the world which would all realize the same chemical state — that the substance in the test tube is copper sulphate. In that sense to say that the test tube contains copper sulphate is already to describe the world in a way which is highly abstracted from the fundamental physical details.

What the moderate reductionist maintains is that we cannot do without causal explanations pitched using abstracted higher-level concepts like 'copper sulphate'. No causal explanations using the concepts of fundamental physics can replicate the explanatory power of abstract causal explanations. Recent work in philosophy of science has helped us to understand how this can be so. Explanation is about answering questions, and a good answer to a why-question includes all relevant information — but only relevant information. Using abstracted kinds enables us to identify causes and effects which are suitably matched to one another in the broad sense that the cause includes all and only the information which is relevant, or *difference-making*, for the effect. A full microphysical description contains a lot of redundant information if all we are interested in is explaining a high-level phenomenon and do not care about the fine details of how that phenomenon is instantiated on a given occasion.

Moving beyond causal explanations, candidates for abstract non-causal explanations take various forms including equilibrium explanations (of why a gas spreads out to fill a container evenly), mathematical explanations (of why periodic cicadas wait a prime number of seasons before emerging), evolutionary explanations (of why male peacocks have such elaborate and unwieldy tailfeathers) and statistical explanations (of why marbles entering a Galton board settle into the shape of a normal distribution). None of the concepts involved in these non-causal explanations are drawn from fundamental physics, and they cannot be replaced with definitions in terms of the concepts of fundamental physics without completely changing the explanatory content of the explanation.

One way to understand why not all explanations reduce to those of fundamental physics, following Yablo (1992) and Menzies and List (2009) is to focus on the need for the explanatory answer to be *proportional* to the corresponding why-question: a proportional answer to a why-question should include all and only the difference-making factors. A related idea is also implemented by Strevens (2008) in the context of his 'kairetic theory of causal explanation'.

If the phenomenon we want explained is specified in higher-level terms, then more often than not the most proportional explanation will also be specified in higher-level terms. Since explanatory proportionality can be characterized rigorously without any reference to the subjective judgments of an agent, the proportionality approach underwrites an objective sense in which the higher-level causes and laws cannot be eliminated in favour of the causes and laws of fundamental physics. It is not just a matter of preference or convenience that we explain phenomena around us in higher-level terms — our use of higher-level abstract explanations is instead understood in terms of tracking what is objectively explanatorily proportional to what.

5. Against Physics Imperialism

The approach to vindicating higher-level explanations and laws outlined in the previous section relies on the idea that abstract higher-level concepts — like molecule, or organism, or belief — could play an indispensable explanatory role by featuring in the most proportionate explanations of the phenomena of interest to us. What, though, of the reductionist argument that since everything is ultimately made of the entities described by fundamental physics, the *real* explanation (never mind the most proportionate) is always to be found at the level of fundamental physics? In this section I will argue that this reductionist argument fails, because fundamental physics cannot *even in principle* be used to model and explain higher-level phenomena such as organisms or beliefs — or even atoms, in complete detail — because of the deep computational intractability of these applications. Physics itself has already uncovered mathematical features of phenomena at the microscopic level which provide the basis for a compelling argument that physics can never be the whole explanatory story concerning phenomena at the macroscopic level. In any world of comparable complexity to the one we inhabit, no explanations in terms of fundamental physics can ever replace the abstract explanations provided by higher-level sciences.

Physics is mathematized right to its core. David Gross's chapter in this volume invokes Galileo's eloquent line that the book of the universe is written in the language of mathematics (Galileo 1623), and the mathematization of physics has only intensified and accelerated since Galileo was writing. There is very little that is common ground across all of contemporary philosophy of physics, but one element of near-universal agreement is that physical theories should be taken and interpreted as they come, in mathematical form, rather than paraphrased into some other language — whether that be English or first-order predicate logic.

The specific details of the mathematization of physics place intrinsic limits on what we can calculate and explain. Even a simple problem like the mutual gravitational interaction of three bodies is not analytically soluble. The limits of mathematics place hard limits on what we can predict and explain about the physical world, given what we have learned about the mathematical structures instantiated by the world. So, ironically, it is the success of mathematical physics itself which leads us inexorably to the conclusion that there are many explanations which mathematical physics will never be able to provide, as a matter not of mere practicality but as a matter of deep mathematical principle.

It is often said that the limitations on using fundamental physics to explain phenomena like weather patterns derive from mere computational limitations: a big enough computer could take the fundamental laws which physicists will eventually identify and use these laws to describe the emergent levels of reality precisely and in all details. The deep limitation to this metaphor is that computation itself is a physical process, and in consequence there are tight constraints imposed by the fundamental laws of physics on what sorts of computations can be performed by a given system within a given timeframe. Even if we say we are interested in what is possible *in principle*, from the point of view of the committed physicalist reductionist the relevant principles will presumably be physical principles rather than philosophical or metaphysical principles. Without delving into the details of analytic solubility, computational tractability and complexity, and other relevant technicalities, we can therefore think in terms of what it would take to simulate the universe using physically possible processes.

In order for us to be in a position to use fundamental physics to answer any possible question whatsoever — whether the question is about sunsets or stars or starlings — notice that we would need to be able to simulate the universe in complete detail. Any gaps in the microphysical description would leave unanswered questions concerning the higher level. We can then appeal to the premise — uncontroversial in the study of the physical basis of computation — that any computer which could simulate some given physical system needs to be many orders of magnitude more complex than the system it attempts to simulate. To attempt to model even the physics of a single proton in terms of its constituent quarks can for mathematical reasons only be done to finite degrees of approximation and even then requires computational resources which will (with current technology) need to be composed of matter containing at least in the order of 10^{27} protons. And things will get a lot more complex as soon as we move beyond an isolated proton. Any computer capable of modelling even a drop of water in something approaching full physical detail would necessarily consume more material and energetic resources than are contained in the entire physical universe.

It is a matter not of the limitations of our intellect but of hard physical law that we cannot use the fundamental laws of physics to simulate any complex phenomena. Instead, we understand the higher-level by patching together relative understandings: we understand features of protons in terms of quarks, atoms in terms of protons, chemistry in terms of atoms, and so on. At each stage, there is no perfect simulation, merely enough explanatory links for us to achieve complete confidence that our theories of the lower level make probable behavior of the general kind described by the higher-level theory. We never have complete derivability of one level from another, in the sense of perfect simulation of the higher in terms of the lower, but we do have a rich network of intertheoretic explanatory notions to draw on which we can use to understand how the lower-level phenomena make the higher-level phenomena possible.

6. Lawless Disciplines

Up to this point I have focused my attention on natural science. But any adequate account of human knowledge must take account of history, anthropology, musicology, and a host of other disciplines which cannot easily be fitted into the model of a natural science. In particular, talk of laws is either minimal or altogether absent — there is a long tradition among historians, eloquently voiced by Patrick Geary in his contribution to the ICA and to this volume, which denies that there are any laws whatsoever in history. Anthropologists frequently think the same way. In this section I will generalize impressionistically and talk of a group of *anarchic* disciplines (often from the humanities) which ignore or explicitly reject the concept of law. What I suggest characterizes many of these disciplines is that they identify the rich interest of the subject-matter as springing from the unique features of the individual cases on which they focus. The understanding of the individual case is enhanced by contextualizing it, but it is actively harmed by the attempt to generalize.

The modalized generality view of laws treats laws as regularities which hold over actual and possible instances of some feature. Even this simple expansion of the focus to contrast an actual case with alternative possible cases draws attention away from the actual case: features that the actual case has which other possible cases lack are abstracted away, which is to say that they are ignored. I am not saying that it is impossible to formulate laws of history or anthropology (though past attempts don't seem to have met with any great success); there is a complex and long-standing debate on the possibility of laws of history (see for instance Hempel 1942 and Geary, this volume) which I don't want to adjudicate here. But I do want to suggest that any laws of history or anthropology we might identify would not be of much interest to historians or anthropologists given their broader theoretical and explanatory projects. Laws of history and anthropology are at best theoretically idle in that they don't play any role in the descriptive, explanatory and narrative content of the works of historians and anthropologists.

To deny that laws of history have any role in the practice of history not to say no lawlike explanations are present in history. Every time there is a causal explanation posited by a historian, the metaphysician of causation will expect a law of some kind to mediate that explanation. Those laws might not be laws of history on any grand scale; for example, if a historian says that the assassination of a leader triggered civil unrest, then the laws involved in this causal claim might be laws of political science or of group psychology rather than distinctively laws of history. The point is that history may traffic in causes, where those causal explanations are mediated by laws of nature, without those laws themselves being laws of history. By way of analogy, biologists may employ explanations where laws of chemistry or physics have important roles to play. To say that an elevated nesting position raises the chance of eggs being lost through falling is not to make the law of gravitational attraction into a law of biology.

In sum, this section has argued that anarchic disciplines exist, that they do not involve any explanatory role for laws of their own, but that they do make use of laws of scientific disciplines to give law-based explanations as part of their broader explanatory and narrative aims. Anarchic disciplines exploit the laws of others, but are themselves subject to none.

7. Conclusion: Laws about Laws

I have surveyed, from a philosophical perspective, the roles of laws in scientific and anarchic disciplines. I have argued that in domains such as biology where laws retain important explanatory power, as well as in more anarchic domains such as history, physics is not and cannot be the whole story — and physics itself is part of the explanation for why that is so.

Can we draw on this discussion to formulate any laws that are about laws? Given the simplicity of the characterization of laws given above, the answer is yes. I have been attempting to say informative things which are true of all possible instances of the concept of a law — accordingly, most of the claims of this article, if they are correct, will themselves count as laws. That is not to say that what I have been doing here is especially deep, or that it is scientific. But it does highlight an underlying commonality between the explanatory role of science and the explanatory role of philosophy.

Acknowledgments

I am grateful to all the organizers and participants of the ICA 3 event for many interesting discussions. Special thanks go to Patrick Geary, for his endless patience with our attempts at interdisciplinary communication, and for his very generous advice and support. This work forms part of the project A Framework for Metaphysical Explanation in Physics (FraMEPhys), which received funding from the European Research Council (ERC) under the European Union's Horizon 2020 research and innovation programme (grant agreement no. 757295). Funding was also provided by the Australian Research Council (grant agreement no. DP180100105).

References

1. B. Carr and B. Ellis, *Universe or Multiverse?* (Cambridge University Press, Cambridge, 2007).
2. A. Eddington, *The Nature of the Physical World* (Cambridge University Press, Cambridge, 1928).
3. A. Einstein, "Time, Space, and Gravitation", *The Times* (London), 28 November 1919, 13–14. Reprinted as "What is the Theory of Relativity?" in A. Einstein, *Ideas and Opinions* (Bonanza Books, New York, 1954).
4. R. Feynman, *The Character of Physical Law* (BBC, London, 1965).
5. Galileo, *The Assayer* (1623).
6. P. Geary (this volume).
7. D. Gross (this volume).
8. S. Green and O. Wolkenhauer, "Tracing organizing principles: Learning from the history of systems biology", *Hist Philos Life Sci* **35**(4), 553–576 (2013).
9. C. G. Hempel, "The Function of General Laws in History", *J. Philosophy* **39**(2), 35–48 (1942).
10. M. Lange, "There sweep great general principles which all the laws seem to follow", *Oxford Studies in Metaphysics* **7**, 154–186 (2012).
11. D. Lewis, "New work for a theory of universals", *Australasian J. Philosophy* **61**(4), 343–377 (1983).
12. J. Leslie, *Universes* (Routledge, London, 1989).
13. C. List, "Levels: Descriptive, explanatory, and ontological", *Noûs* **53**(4), 852–883 (2019).
14. I. Newton, *Philosophiæ Naturalis Principia Mathematica* (1st edition) (1687).
15. P. Menzies and C. List, "Nonreductive physicalism and the limits of the exclusion principle". *J. Philosophy* **106**(9), 475–502 (2009).
16. M. Strevens, *Depth: An Account of Scientific Explanation* (Harvard University Press, Cambridge, 2008).
17. B. Van Fraassen, *Laws and Symmetry* (Oxford University Press, New York, 1989).
18. S. Weinberg, *Dreams of a Final Theory: The Scientist's Search for the Ultimate Laws of Nature* (Random House, New York, 1992).
19. A. Wilson, *The Nature of Contingency: Quantum Physics as Modal Realism* (Oxford University Press, Oxford, 2020).
20. J. Woodward, "Laws: An Invariance-Based Account" in *Laws of Nature*, ed. W. Ott and L. Patton (Oxford University Press, Oxford, 2018).
21. S. Yablo, "Mental causation", *The Philosophical Review* **101**(2), 245–280 (1992).

Chapter 11

Dynamic Cities and Rigid Laws?
Reflections on the Role of Law(s) in Creating Livable Cities[*]

Hanjo Hamann [a], Ulrich Heisserer [b], Nkatha Kabira [c],
Isabel Kusche [d], Irina Kuznetsova [e], Petra Liedl [f],
Tom Schonberg [g] and Carla Aparecida Arena Ventura [h]

[a] *Wiesbaden University of Business and Law (EBS Law School)*
[b] *Avient Protective Materials*
[c] *University of Nairobi*
[d] *University of Bamberg*
[e] *University of Birmingham*
[f] *OTH Regensburg*
[g] *Tel Aviv University*
[h] *University of São Paulo*

Cities play a key role in developing strategies towards making life livable for a large part of the world population and future generations. This chapter explores the potentials and limits of laws to improve livability in cities. Based on an understanding of cities as complex entities, it considers which regulatory tools may be most appropriate to initiate change and what typical barriers they have to deal with. The chapter discusses what laws in the legal sense, the identification and modeling of laws of self-organization as well as the analysis of individual value-based decisions can contribute to a better understanding and governance of continuously evolving cities. It also addresses the entanglement of all governance efforts with informality, the reproduction of class, gender and racial inequalities, and thus questions of social justice. Although there are limitations to legal laws in addressing existing urban injustices due to the idea of legal justice as treating everyone the same, laws nevertheless play a role in making cities livable. They create a framework of rules that limit negative externalities of individuals, which is essential in big agglomerations of people. The challenge is to identify

---

[*] Chapter edited by Isabel Kusche. Author names in alphabetical order.

where such rules are needed and how they may have to be adjusted in view of cities' dynamics.

1. Preface

This chapter aims for a *multi-disciplinary approach* to analyze the interaction of social developments and new technologies related to the role of law in creating livable cities. The authors of this chapter are eight academics from three continents and various disciplines, namely architecture, human geography, engineering, law, public health, sociology and cognitive neuroscience. We are grateful to the organizers of the 3rd UBIAS Intercontinental Academia for bringing us together in Singapore in March 2018 and in Birmingham in March 2019: Prof. Michael Hannon, former Director of the Institute for Advanced Studies at the University of Birmingham, Prof. Eliezer Rabinovici, former director of the Israeli Institute for Advanced Studies from Hebrew University of Jerusalem, Associate Prof. Kwek Leong Chuan, Deputy Director of the Institute of Advanced Studies at Nanyang Technological University of Singapore, Prof. Lars Brink from Chalmers University of Technology and ICA coordinator Sue Gilligan. During the workshops in Birmingham and Singapore we developed the first ideas for this chapter. We also thank Prof. Ernst Rank, Director of the Institute for Advanced Studies at the Technical University of Munich, for hosting us in a very stimulating environment for several days in October 2018 and again in July 2019, during which we discussed and worked on the chapter. The group has jointly reflected on *cities and the law* with the aim of stimulating a discussion about the tension between *rigid laws* and *dynamic cities*. The tension is characterized by the fact that on the one hand, laws are formal and fixed, though not unchangeable, while on the other hand, cities are dynamic, complex in nature and changing all the time. The main question this chapter addresses is: What are the potentials and limits of laws in making cities more livable?

Livability is a "fuzzy concept [...] that means different things to different people but flourishes precisely because of this imprecision".[1] We use the term livability as a placeholder for the characteristics of a city that its inhabitants value and regard as essential for their well-being. This is obviously a very context-dependent or even individual evaluation, which at the same time depends on conditions that are out of control of individuals. That makes the role of laws, which by definition always are somewhat context-independent, in creating livable cities an interesting topic for discussion. The chapter tentatively concludes that there are limits to formal law and as such there needs to be an interplay of top-down approaches stimulating and supporting bottom-up initiatives.

2. Introduction

Global and societal transformation processes determine the 21st century. We experience ecological and social crises, which let societies face enormous challenges. Our current lifestyle endangers the livelihood for future generations. Although urban areas only cover 3% of the earth's land surface[2] they are responsible for 60% to 80% of global greenhouse emissions.[3] Historically, cities concentrated social, political and cultural transformations, including the recognition of women's, ethnic and sexual minorities rights. At the same time, cities can be spaces of poverty, unemployment and segregation and more broadly — social exclusion. With their local innovative capability, cities play a key role in developing strategies towards climate protection and reducing CO_2 emissions.

Cities are complex entities. In its common-sense meaning, complexity means no more than the opposite of simplicity, and if cities were simple, there would be no need for planning departments, urban studies programmes or consultancies for urban design. However, *complexity* is more than a synonym for being *complicated*. Weaver[4] distinguished between problems of *simplicity*, problems of *disorganized complexity* and problems of *organized complexity*. The first are described by very few variables and can therefore be solved with the help of relatively simple equations. The second involve a great number of variables, but since all of them behave in individually erratic ways, these problems can be solved by employing statistics and probability theory. By contrast, problems of organized complexity include a considerable number of variables, which are all interrelated and thus influence one another, leading to *self-organization*.

Cities fall into the category of organized complexity, as do most social phenomena. They change in terms of population size, the businesses and industries supporting their inhabitants, the built environment in which inhabitants live and move, the educational background and aspirations of these inhabitants, and many other respects. Changes in one variable impact many others, and even the rate of change can change as a result of technological innovations, migration flows, environmental problems and other factors. Complex social systems, such as cities, resist planning not only because organized complexity makes prediction of specific events impossible. Cities inevitably create wicked problems,[5] which neither have definitive formulations nor single-best solutions. Consequently, any planning decision is vulnerable to criticism.

In the face of a variety of large-scale, "wicked" problems in systems of organized complexity, laws have often come to be seen as an ineffectual instrument of public governance.[6] City planning and development is one example where visions for the future emphasize flexibility, constant change and the participation of different actors,[7,8] all of which seem to make a recourse to legal regulation outdated or even detrimental.[9] Yet, organized complexity does not mean that cities are best left without any rules and regulations; after all, self-organization is not the same as anarchy.[10]

This raises the question of what kind of regulatory tools may be appropriate for such complex social systems. *Positive coordination*[11] or *teleocracy*[10] is a regulatory approach that aims at specific goals and tries to shape the behavior of various actors in a way that advances these goals. Instruments used to induce the desired behavior include financial incentives or legal prescriptions. By contrast, *negative coordination*[11] or *nomocracy*[10] is a regulatory approach that limits itself to a set of rules that exclude certain interferences and interrelationships between actors but do not prescribe their behavior in the interest of specific goals. In other words, the first is more about self-*organization* and the latter more about *self*-organization.

The two views of regulation have very different implications for the role of laws in cities. Provided the first view, i.e. *teleocracy*, takes complexity seriously, it needs constant feedback on the effects that rules have on actors' behavior and the pursued goals in order to revise them. It needs as much input as possible from various stakeholders as well as the capacity to make sense of it and continuously adjust regulatory tools. Considering that adjustments of rules are likely to impact stakeholders differentially and create a need for new negotiations, it is a huge challenge for such an arrangement to keep up with rapid change. And even if that were possible, the resulting rules would be weak regulatory tools precisely because people would anticipate their being changed again in the near future. Rules that constantly change are difficult to learn and respect.[10] The second view, namely *nomocracy*, by contrast favors laws in the sense of "a stable and simple set of abstract and general relational rules that enable society itself to be highly flexible".[10]

3. Legal Laws and the City

When pondering the role and rule of Law in shaping cities, it is important to note what a city is — and what it is not. Cities exhibit, as we have shown above, emergent behavior and limited central governance. This governance is partly through Law but mostly through other channels of social governance, which we will turn to later. Since Law and social governance are overlapping systems without a clear-cut dividing line, we will narrow the focus for now on a subset of Law, namely formal general Law. This is to say, we consider text which is produced and published by any societal institution with (a) some generally accepted form of moral authority within society, (b) sufficient backing by whoever controls the largest reservoir of violent power in this society, and (c) the intention to influence the behavior of a subset of its people, characterized by abstract properties.

This definition is broad enough to include a variety of institutions across different cultural settings, but our starting point will be promulgated statutory law in a parliamentary democracy. Given this basic understanding then, what is a city *not*? Cities are not generally considered makers of Law. They certainly do have some authority over (certain) local matters, which they exert through ordinances and by-laws. Yet, as the terminology *by*-law suggests,[†] this authority usually (excepting city states such as Hamburg, Hong Kong, and Singapore) derives from higher-order state law. In Germany, for example, formal general law-making is divided without remainder between the 16 German states and the Federal Republic: Art. 70 (1) of Germany's Basic Law reads "The *Länder* shall have the right to legislate insofar as this Basic Law does not confer legislative power on the Federation." Cities, in contrast, are not even mentioned in German constitutional law, except where it determines the capital city to be Berlin. In this sense, cities have no "natural" or "inherent" power. Power is delegated to them by state or federal government, and those delegated powers have been rigorously limited by judicial interpretation. Cities are either subjects of Law, being addressed by parliament as though the city itself was an animate autonomous decision-maker, or they are objects of Law, being merely the quasi-inanimate context wherein animate decision-makers make decisions.

This distinction aligns with a traditional divide in continental law between public Law (being the law in vertical relationships, where the law-maker has authority over the law-subject) and private Law (being the law in horizontal relationships, where two law-makers are each other's law-subjects and bargain over their rights and duties).[‡] This divide means that some formal general Law — the "public" variety — treats "the city" as a subject of "the state" and imposes restrictions on it. For instance, the German state has enacted a law that severely limits the ability of German cities to grow beyond their established boundaries (Sec. 35 Federal Building Code, *BauGB*). It is upon the city's authorities to comply with these limits, lest they be sued in a court of law of their home state and, eventually, the Federal Republic of Germany. The effect of this law is evident to anyone who has ever travelled through Germany and wondered about the extremely compact settlements sprinkled homogeneously across the greenery, with few major metropolises (excepting Berlin and the Ruhrpott) in spite of Germany's wealth, which should exacerbate urban

---

[†] Etymologically, the "by" in by-laws does not derive from a sense of "subordinate" (as in "byway") law, but from an Old Norse word for town, *byr* (https://www.etymonline.com/word/bylaw).

[‡] Note that there is also "private" law in the sense of non-state-enacted law. Its authors rarely command much moral authority except the morality of reciprocity. ("I accept this rule because you accept it.")

sprawl (i.e., a decline of urban density).[12] The law is teleocratic as it shapes the behavior of various actors in a preconceived direction; at the same time it sidesteps the complexity that teleocratic regulation usually entails by insisting on the past as the precedent of what is acceptable in the present and future.

Other formal general Law — the "private" variety — treats the city as a context in which autonomous individuals pursue their well-being and need to be constrained in the interest of other city-dwellers' well-being. Its regulatory effect is nomocratic. For instance, Germany has a law that forces an owner of real estate to accept her neighbor's emissions ("gases, steam, smells, smoke, soot, warmth, noise, vibrations and similar influences") if they comply with the officially designated limits (Sec. 906 German Civil Code, *BGB*). It is upon one real estate owner to discipline the other, with "the city" serving as no more than a backdrop to the story. It is an assumed backdrop (since sub-limit immissions[§] will hardly bother neighbors in the countryside) but no more than that: The city is not treated as an autonomous agent in this relationship.

We therefore see, as was proposed earlier, that cities are not generally considered makers of Law. This does not deny that there actually is some law in the city which is enacted by city officials themselves (municipal law). It may be formal and general, but is usually derivative, i.e., enacted upon the authority granted — or left — to the city by the state. For instance, municipal authorities are allowed to levy some local taxes on businesses, determine the rules for seasonal festivals, or restrict access to public parks. All of these law-making powers are ultimately justified through a "chain of democratic legitimation" which conceptualizes city officials as proxies for state authority rather than truly autonomous rulers over their municipality. They may govern only within the limits and through the leeway left by the state, and their law may not even qualify as Law under the definition set out above, since cities do not wield violent power: The police force and military usually belong to the state, not the city.[**] At the same time, the city is a space where the implementation of

---

[§] While a substance is emitted (released) at the source it is imitted (entering) to another place.
[**] Although all of this varies across jurisdictions. For instance, most police departments in the US are genuine city agencies.

law and law enforcement happen. It is a space where the violation of law and human rights take place as well. The city can be a city of sanctuary and of injustice at the same time.

This should leave lawyers with a puzzling conundrum: If cities are neither autonomous law-makers nor holders of violent power, then why is "the city" nonetheless widely recognized as a significant actor and, in urban studies, as a central unit of analysis? How does the whole (city) become more than the sum of its parts (houses on the territory of an autonomous state)? If it is not just Law in the formal legal sense, then which other "laws" do city-dwellers command — or submit to?

4. Modelling the City

Thinking about cities in terms of complexity and self-organization suggests to look beyond formal law for rules and regularities that govern the city. Since a number of phenomena in physics and chemistry also display characteristics of self-organization, the notion of complexity inspired attempts to unify a thinking in terms of systems and their emergent properties across various disciplines.[13,14] These attempts fundamentally agreed on a number of implications of organized complexity, such as the impossibility of making specific predictions about single states of a system or its elements. Nevertheless, no theoretical consensus emerged on the definition of (organized) complexity or its relevance for understanding the dynamics of social systems in general and cities in particular.

This is not surprising once the notion of complexity is seen as a term that is used to *describe* an entity. It presupposes a distinction between elements and the relations between these elements, but it is indifferent as to how an observer may determine elements and relations.[15] This indifference is open to specification based on disciplinary preferences and habits. Physicists working on complex systems regard cities as examples on which to test general models of complex behavior and properties associated with complexity. Sometimes the idea is to discover surprisingly simple laws that govern various elements and their interrelation in cities. One example is the proposition that many urban measures are predicted by universal scaling laws.[16] It posits both sublinear and superlinear correlations between a city's population size and the size of various urban indicators. For example, infrastructure and services are supposed to scale *sub*linearly due to economies of scale. These economies of scale are regarded as the reason for which cities are able to grow. At the same time, effects from social interaction, such as income or the number of patents, are predicted to grow *super*linearly with population size.

Theoretical physicists such as Geoffrey West thus conceive a city as a sprawling organism that is defined by its infrastructure.[17-19] West has studied examples of cities in the United States, in Portugal and in the UK. One of Geoffrey West's major arguments is that the pace of social life in the cities increases with the increase in the population size in the cities. West understands law, in the scientific sense, as general principles — power laws and scaling laws — that can be tested empirically. West recognizes that, regardless of where a city is located, all cities are governed by 'laws', both human-made and natural. Such laws are emergent properties of underlying structures and act as constraints for cities as systems. In addition, conventions that are rooted in traditions, cultures and other unwritten codes of conduct constrain the system as well. In the long run, laws and other rules are expected to evolve according to the needs of the members of society.

Although the general applicability of scaling laws for cities is still a contested claim,[20] the idea is clear enough: A scaling analysis suggests that the insights into the non-linearity of complex systems can be the starting point for a search for general nonlinear effects, which can be modelled to predict measurable properties of cities as expected averages, in the same way as properties of biological organisms and ecosystems. In this context, population size is not a causal force but a proxy aggregate variable that stands for general effects of intense interactions between co-located people.

In the same spirit of quantification, other approaches make use of computer simulations to grasp the organized complexity of cities and model disequilibria and dynamics according to ideas taken from physics.[14] For example, once similarities between spatial regularities in cities and fractals, i.e. objects with self-similar form at different scales, such as snowflakes, had been recognized, algorithms for generating fractals became the foundation for simulating certain spatial properties of cities.[14]

The approaches guided by physics share a view of cities that treats them like any other complex system. An obvious objection is that there is an essential difference between natural entities like snowflakes or ecosystems and cities, which is their being the result of human culture, intentions, norms and policies.[21] Consequently, interrelations between elements, however these are defined, are infused with meaning and in that sense very different from natural entities. This suggests a slightly different take on complexity, which stresses the difference between systems in which all elements can still be related to each other at the same time and systems with so many elements that their concatenation is inevitably selective.[15] Selectivity implies contingency, i.e. it makes visible that other selective concatenations would be possible as well. Therefore, there is an inherent connection between organized complexity and decision-making. Both individuals and organizations constantly take decisions related to the city. The two following sections propose perspectives for thinking about their respective roles and capturing their contribution to cities' complexity.

5. Individual Decision-Making in the City

As discussed above, cities are highly complex environments, comprising multiple layers of stationary and moving components. Humans live within the city and act as dynamic agents with multiple interactions.[22] One of the most important behaviors of all living organisms is decision-making. As humans we probably make tens of individual decisions every day. Many of these decisions can be regarded as value-based decisions, where the benefits and costs are weighed until a decision is reached. The process by which animals and humans perform individual value-based decisions has been studied extensively over the past two decades.[23] City dwellers need to behave within the environment with acceptable laws. It is thus an intriguing question what role individual decision-making plays in the creation of the complexity of the city.

Laboratory studies of decision-making processes can aid in answering this 'real world' question. Some individual decisions with huge aggregate consequences for cities happen infrequently, for example the choice of where to live. This type of decision indeed involves many factors related to benefits and costs, such as the ability to purchase a larger property in a less expensive, but more violent neighborhood. However, as this is a relatively rare decision in an individual's life, the problems with quantifying the different components of such a decision, inferring individual parameters and predicting the future behavior of an individual are almost insurmountable. Rangel *et al.*[23] propose a computational learning framework that can be applied to reinforcement-learning paradigms, which are decisions that are repeated over and over again and thus allow optimizing individual parameters. An example for such a value-based decision is commuting and transportation within the city.

Optimizing transportation in a city requires that planners consider the city structure and its inhabitants' needs. The individual value-based decision-making concept used within this realm is the Value of Travel Time. Planners used to mainly focus on the economic aspect of time saving.[24] However, nowadays there is a realization that other aspects go into each individual's decision on transportation that can affect the entire system. For example, a project funded by the EU from 2017 to 2020, named "Mobility and Time Value" (MoTiV), provides a novel perspective on transportation by changing the orthodox view of optimizing the Travel Time Budget (TTB) that each individual allocates to their daily commute. The approach that MoTiV proposes takes into account multiple factors of individual well-being[25] to *wasted* or *worthwhile* time. It weighs various benefits and costs beyond time and money.[26] The project used a specially designed app named *woorti,* which collects data of commuters along three scales of "productivity", "enjoyment" and "fitness". The project leaders suggest that these data will allow city planners to consider richer individual-level data that go beyond the aspect of travel time when they further develop and optimize transportation systems.

Yet, transportation systems also change from below, without any initial planning, and individual behavior in transportation does not always adhere to existing rules and laws. It can occasionally act outside them and lead to a realization that new laws need to be drafted. An example are e-bikes or e-scooters as new modes of transportation. They have become very common in densely populated cities around the globe as they offer a highly efficient and cheap individual transportation. However, when these modes of transportation are not regulated, they may disturb the balance between individual utility and general public needs. New laws of the nomocratic type are needed to regulate the use of these vehicles and to ensure that the balance between the rights of different commuters in the city is preserved and optimized.

Figure 1: Signs reminding commuters in Tel Aviv of new fines given to e-scooters and e-bike riders if caught on sidewalks.

An example of such new regulations regarding e-scooters and e-bikes are those imposed by the Tel Aviv municipality. These vehicles are not allowed on sidewalks (see Fig. 1) but on bike lanes and the roads. The introduction of new fines may change the delicate ecosystem of usage of different modes of transportation, since riders will need to incorporate the risk of an accident on the road versus the risk of a fine into their value-based decisions. Interestingly, since these vehicles do not have a license plate, it is hard to keep track of them and actually impose the fine. The use of e-scooters thus illustrates the limited power that cities have as well as how they may use the power and ask the government to come up with new laws at the same time.[27]

Individual transportation is a prime example of how value-based decision-making theories, based on laboratory experiments, may be applied to real world examples and could contribute to creating laws and regulations that promote better well-being. Usage of optimization tools that take into account individual benefits could offer each individual an optimal mode of transportation, while cities and countries would need to create laws that adapt to these new transportation modes.

E-bikes and e-scooters are an example of low-emission, affordable individual modes of transportation. Hence they are often used by the lower income population as their investment costs are lower compared to cars and they do not require parking nor high maintenance. In Tel Aviv, they are in particular used by students and immigrants living in the southern parts of the city to reach their workplaces without being dependent on public transportation, which does not cover all suburbs properly. This is an example of how individual decisions can further social justice in the city by allowing the lower economic strata to move into other richer parts of the city despite a lack of public transportation. Yet, once the vehicles became very widespread calls to limit their usage began and eventually new laws were implemented. This points to the conflictual dimension of complexity in cities and the key role that questions of social justice play when it comes to their livability.

6. Social Justice, Difference and the Right to the City

On the one hand, individual decision-making doubtlessly shapes mobility patterns and many other social practices in the city. On the other hand, such patterns and practices are also shaped by the city and its complex formal and informal regulations. In addition to individuals, various organizations, including city councils and private companies, are involved in this mutual structuration.[28] At the same time, individuals can also be considered in their heterogeneity instead of conceiving them as value-based decision-makers. This introduces a critical perspective on cities and their relation to social justice.

> "Corporate producers of space tend to define the public as passive, receptive and refined, fostering the illusion of a homogenized public, by filtering out the social heterogeneity of the urban crowd, with minimal exposure to the horrifying level of homelessness and racialized poverty that characterizes the street environment".[29]

Urban space is highly political and ideological, where law and lawlessness, formal order, and fluid informality happen. As politicized and ideological space, the city supports and reproduces class, gender, and racial inequalities. In this respect, it is where the disparities of normative rules implemented by law and everyday life take place.

The understanding of urban space and society as mutually constitutive is central to the notion of the right to the city.[30] Its revolutionary vision of the production of space, introduced by Lefebvre[31] in the book that took this notion as its title, is based on the Marxist analysis of production and articulates space as "not a scientific object removed from ideology or politics. It has always been political and strategic as there is an ideology of space. Because space, which seems homogeneous, which appears as a whole in its objectivity, in its pure form, such as we determine it, is a social product".[31] The idea of the right to the city became a flagship of the critical geography in the 20th century in its effort to overcome the alienation from the city produced by control of the state and capitalism.[32–34]

The city is "a man's most consistent and on the world, his most successful attempt to remake the world he lives in more after his heart's desire". Thus, the city is the world which man created, and, at the same time, the world he is condemned to live in. Therefore, the right to the city is more than an individual right to access resources, it represents the right to change ourselves by changing the city, which depends upon the exercise of a collective power over the process of urbanization.[34] The definition of "right to the city" aims to highlight the non-exclusion of any part of the society from access to urban life quality and benefits. The claim for the right to the city expresses issues related to urban development and the effects of political and economic crises. This claim demands a higher degree of democratization in the cities and more collective decision-making processes based on the principles of solidarity, freedom, equity, dignity, and social justice.[35]

The concept of the right to the city is founded on the ethics of human rights, as initially defined in the UN Declaration, but does not form part of the human rights regime. It has been embedded in program documents of UN-HABITAT and UNESCO, and it also influenced the World Charter for the Right to the City, the European Charter for Human Rights and the City, as well as the Montreal Charter of Rights and Responsibilities. It has become an inspiration for communities across many countries.[36] Nevertheless, it is still considered a precarious and neglected human right.[34] As David Harvey argued, "To claim the right to the city (...) is to claim some kind of shaping power over the processes of urbanization, over the ways in which our cities are made and re-made and to do so in a fundamental and radical way."[37]

One of the criticisms of Lefebvre's approach to the right to the city was his primary focus on the working class. Purcell[32] suggests to go beyond it with a "variegated politics of identity and difference". That continues the seminal conceptual framework for the politics of difference developed by Iris Young,[38] who argued that "social justice in the city requires the realization of a politics of difference. These politics lays down institutional and ideological means for recognizing and affirming diverse social groups by giving political representations to these groups". Young[38] moves away from the redistribution mode of welfare capitalism and articulates five categories of oppression that can be applied to any group, namely exploitation, marginalization, powerlessness, cultural imperialism, and violence. If Young[38] focuses on social groups as objects of oppression, Harvey[34] adds the nature in cities, as it also has rights, stressing that "Just planning and policy practices will clearly recognize that the necessary ecological consequences of all social projects have impacts on future generations as well as upon distant peoples and take steps to ensure a reasonable mitigation of negative impacts".[34]

A recent turn in the "right to the city" approach suggests to consider not only aspirational rights but also rights which are "informally negotiated between communities (or individual actors) in the streets".[39] It is therefore necessary to distinguish between formal citizenship within the nation-state, and the exercise of urban citizenship through democratic practice. Substantive practices of citizenship emphasize the difference between rights and the ability to enjoy and perform such rights. Substantive citizenship is acquired through[40] participation and enacted through participatory democracy.[41] Substantive citizenship can be exercised at several levels, one of which is the city. The right to the city signifies a societal ethics cultivated through living together and sharing urban space. It concerns public participation, where urban dwellers possess rights and cities — city governments and administrations — possess obligations or responsibilities. Civil and political rights are fundamental, protecting the ability of people to participate in politics and decision making by expressing views, protesting and voting. The exercise of substantive urban citizenship thus requires an urban government and administration that respects and promotes a societal ethics. It also demands responsibilities of citizens to use and access the participatory and democratic processes offered.[42] The formulation and materialization of a new political contract of social citizenship, recognizing and legalizing the rights of citizens to participate fully and actively in political and civil society form the *sine qua non* condition for the expansion and deepening of democracy. Such a widening of citizenship rights becomes even more important for the promotion of democratic governance of cities: cities and citizenship are ultimately the same subject.[43]

7. Communities, Laws and Urban Planning

Public participation in urban planning is one of the key instruments of social citizenship in a city. So to what extent is law required as an instrument for the creation of livable urban space? Abdou Maliq Simone,[44] who studied informal settlements in countries in the Global South for three decades, emphasizes the role of creativity and self-organization which are far removed from formal laws. He argues that in informal settlements "provisionality is being engaged as the pretext for elaborating engagements with the urban that seek protracted opportunities for experimenting with livelihood, territorial emplacement, and domestic organization. Particular ways of seeing, believing, and knowing accompany these experiments, which residents themselves frequently sum up as paying attention to the background". Simone[45] describes urban everyday life as "popular economy", where "informality is intensely situated in the specifics of all kinds of articulations and imbrications in many other processes".[46]

Similarly, the idea of a city as a lived space with its spontaneity and diversity is a core of Jane Jacobs's work,[47–50] although she stresses the role of human rights and the necessity to struggle for urban change. Her focus on street-level self-organization also resonates with the notion of rights that are informally negotiated between individuals or communities in the city. As an urbanist, Janet Jacobs conceived cities as the prime drivers of economic development.[48] She was the founder of Vancouverism, a city planning technique that is characterized by medium-height commercial base, and arrow high rise residential towers. She relied on the notion that cities evolve and focused on generators of diversity.[51] Jacobs saw cities as integrated systems that have their independent dynamism and logic, which changes over time depending on how they are used. For Jacobs, experience, observation and culture are critical in understanding and analyzing the city. She studied cities such as Toronto, Philadelphia, Quebec and New York, especially Manhattan and more specifically Greenwich. Her key message is that urban designing cannot be carried out in boardrooms, as urban centers are not abstract.[47] Jacobs[47] argues that streets are the lifeblood of urban centers and that neighborhoods should be capable of serving several functions so that people may be on the streets all the time. Furthermore, she argues that buildings and intricate street structures are of benefit to cities. She also advocates for a high degree of concentration of people. Jacobs further points out that cities are the driving force for development and prosperity, and private investment shapes the city, but private investment is shaped by laws. In this sense, she considers the role of private law in constraining nomocratically (see Introduction) the ways in which autonomous individuals can pursue their goals in the context of cities. At the same time, she emphasizes that such law is a product of social construction, which implies that it could be changed and that the cities emerging from the conduct of individuals would change to some degree with it, for better or worse.

In a similar vein, Kelvin Campbell[52–54] argues that urbanism in the sense of a good way of life for cities' inhabitants is only delivered through effective economic change and the building of cultural capital. He uses what he calls "massive small techniques" to analyze examples of urban change in Cape Town, London, Johannesburg and Rio de Janeiro. Campbell argues that the collective power of many small ideas and actions can make a big difference in the process of creating an ideal urban center.[54] He argues for collective or rather collaborative action by the government and the citizens in the process of building cities. According to Campbell, laws are responsible for creating conditions for better urban centers. Small laws and rules generate exemplarily complex systems. Campbell acknowledges that the laws governing cities may be written or unwritten. For him, laws, codes, and by-laws shape smart cities' agendas, with 'smartness' taking on a wider meaning that goes beyond or sidesteps its mainstream understanding of the datafied city. Furthermore, laws have a controlling effect on urban structure. In addition, laws create the conditions necessary for the formation of a neighborhood. He considers laws as tools of change but argues against complex policies as he considers them rigid and arrestive. In that sense, he trusts negative coordination or nomocracy and is highly skeptical of attempts at positive coordination or teleocracy (see Introduction).

Other studies also suggest that laws have a role in city planning but that there are also limits to addressing socio-spatial justice through legal tools. The example of Massachusetts' equal access laws in the realm of social service provision, which is considered to be one of the frontiers of urban justice referred to as 'the Dover Amendment', indicates the limits of the law for urban justice efforts. These limits are a result of the economic, social, and political context and historically framed landscapes of poverty in a region.[39,55] On the basis of their research of that case, Pierce and Martin[55] suggest four propositions about the justice potential of the law at the urban scale: 1) "legal tools can support agents working towards urban justice by shielding them from state interference", but "it is never sufficient for urban justice"; 2) "legal tools cannot enable substantive urban justice without also redistributing concomitant social and economic resources to support those outcomes" 3) "legal tools offer little leverage to those seeking systematic revolution except via cynical, strategic abuse of the state"; 4) "the law may be used tactically in concert with other, non-state or non-legal tools towards justice outcomes".

These limitations of legal tools can be taken into account and mitigated to some degree in the case of democratic countries that support free and fair elections, strong and independent institutions, and political rights, such as the right to protest, and civil rights like access to a fair trial. However, considering that most countries in the world have flawed democracies, hybrid or autocratic regimes, the barriers to achieving social justice in cities at global scale are very high.

Other limitations are closely intertwined with the question of 'whose rights and whose city'. They are related to the neoliberal logic of urban development in many regions, and the lack of resources of those who suffer from the inequalities for the advocacy and struggles for their rights. Looked at economically, the cry for the Right to the City here comes from the most marginalized and the most underpaid and insecure members of the working class, not from most of the gentry, the intelligentsia, the capitalists.[56] Despite the assumption that urbanization is followed by the strengthening of women's rights, the women in the Global South do not benefit much from the growing prosperity.[57]

As Robinson[58] stresses, urban theory should acknowledge the differences between experiences of post-colonial cities and wealthy big cities. The 'ordinary' cities in all their complexity and diversity have to be central for research and policies. For example, "traditional western industrializing urbanization accompanied by the rise of middle classes is hardly happening in much of Africa".[59] Instead, in many cities, urbanization is characterized by a large informal and "survival-oriented economic activity", which reinforces an "externally dependant character of urbanization". Some cities in Eastern Europe found themselves in pitfalls of hybrid spatialities that emerged from the mutual embeddedness of neoliberalism and socialist legacy,[60] which means that the spatial organization of the city and social and economic relations impact on the potential for urban change.

8. Conclusion

So what are the potentials and the limits of laws in making cities more livable? Much of what we discussed in this chapter seems to foreground the limits. First of all, cities are not autonomous lawmakers, although they are undoubtedly significant actors and widely recognized as central units of analysis. Furthermore, cities are characterized by organized complexity, which implies a self-organization that is not easily reconciled with laws, especially the type of laws that aims at furthering specific public goals, however desirable they may seem in the abstract.

Since the self-organization of cities involves conscious agents who ascribe meaning to their conduct and decisions, we also have to consider the role of individual decision-making and the role of conflicts around social justice in the city. As we have shown with the help of the e-scooter example, individuals' value-based decisions are related to laws in several ways. They will take them into account when making decisions about the means of transport to use, but will not necessarily obey them. Their individual decisions, once they are better understood, for example with the help of lab experiments, may in the aggregate lead planners and lawmakers to consider new rules in the interest of improving a city's transportation infrastructure. Yet, there is no pre-established harmony when it comes to determining whether a new rule is needed and what it should look like. Instead, a city's heterogeneities in terms of class, gender and ethnicity may lead to manifest conflict or to denying a part of its inhabitants a life in the city that furthers their well-being.

Through the lens of social justice, we found that the diversity of social, political and economic inequalities impact upon cities' vulnerabilities and puzzle the potential of laws to address multiple urban injustices. Legal laws are bad at addressing pre-existing inequalities since the idea of legal justice is based on treating everyone equally even though they are not equal due to the role of class, ethnicity, race, gender and capitalist production.

Yet, we found that laws nevertheless play a role in making cities livable by creating a framework of rules that limit negative externalities of individuals for other people. It is difficult to imagine big agglomerations of people without such basic rules; the challenge is to identify where such rules are needed and how they may have to be adjusted in view of cities' dynamics, for example in adopting new technologies like the e-scooter. It is in the absence of such rules that informal rules inevitably emerge. Although they play a role in all cities, they are especially central in settings in which the rule of law is not guaranteed. This aspect finally leads back to the city as an entity that is not an autonomous lawmaker. Despite the seeming rigidity of many laws in the context of fast-changing cities, a self-organization of cities that promotes livability for all at a minimum requires a reliable framework of laws ensuring fundamental rights of their citizens.

We conclude that, to make cities more livable, formal laws are needed as part of an interplay in which top-down approaches set the framework to stimulate and support bottom-up initiatives and change.

References

1. Markusen A. Fuzzy concepts, proxy data: Why indicators would not track creative placemaking success. *Int J Urban Sci* [Internet]. 2013 Nov; 17(3):291–303. p. 293. Available from: http://www.tandfonline.com/doi/abs/10.1080/12265934.2013.836291
2. University C for IESIN-C-C, IFPRI IFPRI-, Bank TW, CIAT CI de AT-. Global Rural-Urban Mapping Project, Version 1 (GRUMPv1): *Land and Geographic Unit Area Grids* [Internet]. Palisades, NY: NASA Socioeconomic Data and Applications Center (SEDAC); 2011. Available from: https://doi.org/10.7927/H48050JH
3. Kamal-Chaoui L, Robert A, editors. *Competitive Cities and Climate Change.* 2nd ed. OECD Regional Development Working Papers, OECD publishing, © OECD; 2009. 172 p.
4. Weaver W. Science and complexity. *Am Sci.* 1948;36(4):536–544.
5. Rittel HWJ, Webber MM. Dilemmas in a general theory of planning. *Policy Sci* [Internet]. 1973 Jun;4(2):155–69. Available from: http://link.springer.com/10.1007/BF01405730
6. Benz A, Papadopoulos I. *Governance and Democracy: Comparing National, European and International Experiences.* Routledge; 2006. 272 p.
7. Newman P. *Sustainable Cities of the Future: The Behavior Change Driver. Sustain Dev Law Policy.* 2010;11(Issue 1 Fall 2010: Sustainable Development in the Urban Environment):7–10.
8. Campbell K. Smart urbanism: Making massive small change. *J Urban Regen Renew.* 2011;4:304–11.
9. Howard PK. *The Death of Common Sense: How Law Is Suffocating America.* Random House Trade Paperbacks; 2011. 256 p.
10. Moroni S. Complexity and the inherent limits of explanation and prediction: Urban codes for self-organising cities. *Plan Theory* [Internet]. 2015 Aug 12;14(3):248–67. Available from: http://journals.sagepub.com/doi/10.1177/1473095214521104
11. Scharpf FW. Games Real Actors Could Play. *J Theor Polit* [Internet]. 1994 Jan 29;6(1):27–53. Available from: http://journals.sagepub.com/doi/10.1177/0951692894006001002

12. Patacchini E, Zenou Y, Henderson JV, Epple D. Urban Sprawl in Europe. *Brookings-whart Pap Urban Aff* [Internet]. 2009;125–49. Available from: http://www.jstor.org/stable/25609561

13. Haken H. Complexity and Complexity Theories: Do These Concepts Make Sense? In: *Complexity Theories of Cities Have Come of Age* [Internet]. Berlin, Heidelberg: Springer Berlin Heidelberg; 2012. p. 7–20. Available from: http://link.springer.com/10.1007/978-3-642-24544-2_2

14. Batty M, Marshall S. The Origins of Complexity Theory in Cities and Planning. In: *Complexity Theories of Cities Have Come of Age* [Internet]. Berlin, Heidelberg: Springer Berlin Heidelberg; 2012. p. 21–45. Available from: http://link.springer.com/10.1007/978-3-642-24544-2_3

15. Luhmann N. *Theory of Society, Volume 1*. Stanford University Press; 2012. 488 p.

16. Bettencourt LMA, Lobo J, Strumsky D, West GB. Urban Scaling and Its Deviations: Revealing the Structure of Wealth, Innovation and Crime across Cities. Añel JA, editor. *PLoS One* [Internet]. 2010 Nov 10;5(11):e13541. Available from: https://dx.plos.org/10.1371/journal.pone.0013541

17. Bettencourt L, West G. Bigger Cities Do More with Less. *Sci Am*. 2011;305(3):52–3.

18. Bettencourt LMA, Lobo J, Helbing D, Kuhnert C, West GB. Growth, innovation, scaling, and the pace of life in cities. *Proc Natl Acad Sci* [Internet]. 2007 Apr 24;104(17):7301–6. Available from: http://www.pnas.org/cgi/doi/10.1073/pnas.0610172104

19. West G. *Scale: The universal laws of growth, innovation, sustainability, and the pace of life in organisms, cities, economies, and companies*. Penguin Press; 2017. 496 p.

20. Arcaute E, Hatna E, Ferguson P, Youn H, Johansson A, Batty M. Constructing cities, deconstructing scaling laws. *J R Soc Interface* [Internet]. 2015 Jan 6;12(102):20140745. Available from: https://royalsocietypublishing.org/doi/10.1098/rsif.2014.0745

21. Portugali J. Complexity Theories of Cities: Achievements, Criticisms and Potentials. In: Portugali J, Meyer H, Stolk E, Tan E, editors. *Complexity Theories of Cities Have Come of Age An Overview with Implications to Urban Planning and Design*. Springer Berlin Heidelberg; 2012. p. 47–62.

22. Bonabeau E. Agent-based modeling: Methods and techniques for simulating human systems. *Proc Natl Acad Sci* [Internet]. 2002 May 14;99(Supplement 3):7280–7. Available from: http://www.pnas.org/cgi/doi/10.1073/pnas.082080899

23. Rangel A, Camerer C, Montague PR. A framework for studying the neurobiology of value-based decision making. *Nat Rev Neurosci* [Internet]. 2008 Jul 11;9(7):545–56. Available from: http://www.nature.com/articles/nrn2357

24. Wardman M. The Value of Travel Time: A Review of British Evidence. *J Transp Econ Policy*. 1998;32(3):285–316.

25. University of Žilina S. *About MoTiV* [Internet]. Available from: https://motivproject.eu/about-motiv/objectives.html

26. Scitech Europa Quarterly. *A digital agenda - A concerted effort is needed in order to realise a strong digital future* [Internet]. Available from: http://edition.pagesuite-professional.co.uk/html5/reader/production/default.aspx?pubname=&edid=4bbfddf0-5b47-4a35-b797-80527690bef0

27. Hadar T. *9,000 Tickets Issued to E-Scooter and E-Bike Users in Tel Aviv in 2019* [Internet]. 2019. Available from: https://www.calcalistech.com/ctech/articles/0,7340,L-3763282,00.html

28. Giddens A. *The Constitution of Society: Outline of the Theory of Structuration.* John Wiley & Sons; 2013. 438 p.

29. Crislley D. Megastructures and Urban Change: Aesthetics, ideology and design. In: *The Restless Urgan Landscape.* P Knox. Englewood Cliffs, N.J. : Prentice Hall; 1993. p. 127–64.

30. McCann EJ. Space, citizenship, and the right to the city: A brief overview. *GeoJournal* [Internet]. 2002;58(2/3):77–9. Available from: http://link.springer.com/10.1023/B:GEJO.0000010826.75561.c0

31. Lefebvre H. *The production of space.* Trans. D. Oxford, UK, and Cambridge, MA: Blackwell; 1991.

32. Purcell M. Excavating Lefebvre: The right to the city and its urban politics of the inhabitant. *GeoJournal* [Internet]. 2002;58(2/3):99–108. Available from: http://link.springer.com/10.1023/B:GEJO.0000010829.62237.8f

33. Mitchell D. *The Right to the City: Social Justice and the Fight for Public Space.* The Guilford Press; 2003. 270 p.

34. Harvey D. The right to the city. *Int J Urban Reg Res* [Internet]. 2003 Dec;27(4):939–41. Available from: http://doi.wiley.com/10.1111/j.0309-1317.2003.00492.x

35. Figueiredo GLA, Martins CHG, Damasceno JL, Castro GG de, Mainegra AB, Akerman M. Direito à cidade, direito à saúde: quais interconexões? *Cien Saude Colet* [Internet]. 2017 Dec;22(12):3821–30. Available from: http://www.scielo.br/scielo.php?script=sci_arttext&pid=S1413-81232017021203821&lng=pt&tlng=pt

36. Purcell M. Possible Worlds: Henri Lefebvre and the Right to the City. *J Urban Aff* [Internet]. 2014 Feb;36(1):141–54. Available from: https://www.tandfonline.com/doi/full/10.1111/juaf.12034

37. Harvey D. *The Right to the City* [Internet]. p. 16. Available from: https://davidharvey.org/media/righttothecity.pdf

38. Young I. *Justice and the Politics of Difference.* Princeton University Press; 2011. 304 p.

39. Pierce J, Williams OR, Martin DG. Rights in places: An analytical extension of the right to the city. *Geoforum* [Internet]. 2016 Mar;70:79–88. Available from: https://linkinghub.elsevier.com/retrieve/pii/S0016718516300495

40. Dikeç M, Gilbert L. Right to the City: Homage or a New Societal Ethics? *Capital Nat Social* [Internet]. 2002 Jun;13(2):58–74. Available from: http://www.tandfonline.com/doi/abs/10.1080/10455750208565479

41. Brown A. *Contested Space: Street trading, public space and livelihoods in developing cities.* Rugby, ITDG; 2006.

42. Brown A, Kristiansen A. *Urban policies and the right to the city.* UNESCO; 2008. 56 p.

43. Fernandes E. Constructing the 'Right To the City' in Brazil. *Soc Leg Stud* [Internet]. 2007 Jun 17;16(2):201–19. Available from: http://journals.sagepub.com/doi/10.1177/0964663907076529

44. Simone A. Maximum exposure: Making sense in the background of extensive urbanization. *Environ Plan D Soc Sp* [Internet]. 2019 Jun 18;026377581985635. Available from: http://journals.sagepub.com/doi/10.1177/0263775819856351

45. Simone A. *Improvised Lives: Rhythms of Endurance in an Urban South.* Polity; 2018. 120 p.

46. Simone A. Contests over value: From the informal to the popular. *Urban Stud* [Internet]. 2019 Feb 20;56(3):616–9. Available from: http://journals.sagepub.com/doi/10.1177/0042098018810604

47. Jacobs J. *The death and life of great American cities.* Reissue. Vintage Books; 1961. 480 p.

48. Jacobs J. *The Economy of Cities.* First Pr edition, Vintage; 1970.

49. Jacobs J. *Cities and the Wealth of Nations: Principles of Economic Life.* Reprint. Vintage Books USA; 1985. 257 p.

50. Alexiou A. *Jane Jacobs: Urban Visionary.* Rutgers University Press; 2006. 224 p.

51. Lueders A. 'Exploring the Legacy of the 20th Century's Most Provocative Urban Theorist.' 2016.

52. Campbell K. *By Design, Urban Design in the Planning System: Towards Better Design.* Thomas Telford Publishing; 2000. 104 p.

53. Campbell K. *Re:urbanism: A Challenge to the Urban Summit.* Urban Exchange; 2006.

54. Campbell K. *Massive Small: The Operating System for Smart Urbanism.* Urban Exchange; 2010.

55. Pierce J, Martin D. The law is not enough: Seeking the theoretical 'frontier of urban justice' via legal tools. *Urban Stud* [Internet]. 2017 Feb 20;54(2):456–65. Available from: http://journals.sagepub.com/doi/10.1177/0042098016636574

56. Marcuse P. From critical urban theory to the right to the city. *City* [Internet]. 2009 Jun 2;13(2–3):185–97. Available from: https://www.tandfonline.com/doi/full/10.1080/13604810902982177

57. Chant, S. 2013. Cities through a "gender lens": A golden "urban age" for women in the global South? *Environment and Urbanization* [Internet]. 2013. 25(1): 9–29. Available from: https://doi.org/10.1177/0956247813477809

58. Robinson. *Ordinary cities: Between Modernity and Development.* Routledge; 2006. 198 p.
59. van Noorloos F, Kloosterboer M. Africa's new cities: The contested future of urbanisation. *Urban Studies* [Internet]. 2018 May 24;55(6):1223–41. Available from:
    http://journals.sagepub.com/doi/10.1177/0042098017700574
60. Golubchikov O, Badyina A, Makhrova A. The Hybrid Spatialities of Transition: Capitalism, Legacy and Uneven Urban Economic Restructuring. *Urban Studies* [Internet]. 2014 Mar 15;51(4):617–33. Available from: http://journals.sagepub.com/doi/10.1177/0042098013493022

Chapter 12

ICA 3 Singapore and Birmingham: Administrators' Perspective, Supporting and Delivering a Vision

Leong Chuan Kwek[*] and Sue Gilligan[†]

[*]Centre for Quantum Technologies, National University of Singapore
[†] Institute of Advanced Studies, University of Birmingham, UK

1. Introduction

Royal Shakespeare Company, Stratford-upon-Avon. Standing in a circle, we read words from a passage from "As You Like It" one word at a time:

All the world's a stage,
And all the men and women merely players;
They have their exits and their entrances,
And one man in his time plays many parts,
His acts being seven ages.

We learnt how to say the words with the needed nuances and diction. Nobody was spared. Not even a Nobel Laureate. But what have these exercises got to do with a research theme on "Laws: Rigidity and Dynamics"? Perhaps everything. Just as there are laws governing society, there are laws governing science and there are laws governing diction and play. Fellows and Mentors described this workshop led by Cathleen McCarron, RSC Senior Voice Practitioner and National Theatre Head of Voice, as an intellectual highlight and the most enjoyable cultural activity of the ICA.

This unique opportunity to dissect the text word by word, reflecting on meaning and rhythm, listening to and working with others, resonated with the everyone involved in the ICA project, as another tool to understand collaborations across disciplines and the value of individual voices coming together for coherent purpose.

The Intercontinental Academia represents a long-term commitment for host institution with a complex project comprising two legs: each leg is organized by an Institute of Advanced Studies on a different continent.

The mission for the administrative team, to support the vision, is to:

- fit together the complex jigsaw of inter-generational, continental, disciplinary and cultural elements into a cohesive programme;
- plan a programme and environment which can stimulate, inspire and nurture interactions across our theme;
- deliver events which showcase Institutes to colleagues, students and the public at the host city;
- evaluate and improve the ICA concept to pass the baton on to the next ICA.

ICA 3 Timeline

Jun 2016: UBIAS Directors' Meeting in Birmingham partnership is accepted and funding negotiated.

Sep 2016: The theme is agreed.

Oct 2016: The search for mentors begins.

Jan 2017: The recruitment of fellows begins.

Sep 2017: Fellows selected.

Mar 2018: 10-day meeting, Singapore.

Oct 2018: Interim Project Meeting, Institute of Advanced Study, Princeton (IAS).

Oct 2018: Interim Project Meeting, Technical University of Munich's Institute for Advanced Study (TUM-IAS).

Mar 2019: 10-day meeting, Birmingham.

---

2. The Partnership

Under the Directorship of Professor Eliezer Rabinovici and Professor Phua Kok Khoo, The Institute for Advanced Studies at Nanyang Technological University Singapore organized the first leg in March 2018.

Led by Professor Michael Hannon at the Institute of Advanced Studies, The University Birmingham hosted the second leg a year later in March 2019. Birmingham, the social, financial, scientific, cultural, and commercial hub of the Midlands in UK, is closely connected to Singapore and Professor Phua Kok Khoo is an alumnus of the University of Birmingham.

3. The Theme "Laws: Rigidity and Dynamics"

The notion of laws carries different meanings in different settings, ranging from exact sciences to social sciences to cultures and individuals. 'Laws' is a common underlying thread from Chemistry to Linguistics, History to Physics, Engineering to Economics and beyond. This rich theme was selected as it demands reflection, analysis and comparison by common and distant disciplines, with the aim of finding connections and resonances and developing new dialogues.

4. The Mentors

The Directorate identified and approached eminent mentors. In addition to world class researchers in their field it was essential that the mentors were committed to the aims and purpose of the ICA and the importance of supporting early careers researchers in interdisciplinary dialogue.

Once participation has been agreed, the administrative team are then in close contact on scheduling and the organisation of travel and accommodation. Mentors do not ask for stipends, but it is expected that all practical aspects of their involvement will be taken care of to make the, often long, journeys and participation as comfortable as possible.

5. Recruitment of Fellows

One of the most challenging aspects of the organisation is the recruitment and selection of fellows. Success for ICA demands a group of outstanding early career researchers, usually within 15 years of completing their doctoral research, who have embarked on, and are clearly set to continue successful careers in academia or industrial research. The fellows need to demonstrate an interest in multi-disciplinary research and the ability to challenge and benefit from interactions with our eminent mentor group.

Using the newly established ICA website, active advertisement and promotion started a year before the first meeting through members of the University Based Institutes for Advanced Study (UBIAS), under whose auspices the ICA takes place. Applicants are expected to be sponsored by either an Institute for Advanced Study or their home institution. Sponsors will act as a referee, be able to guarantee that if the fellow is selected that they can be released for two relatively long workshops of 10-day duration and will also cover the costs of travel. The management of the host institutions, ICA Directors and Mentors also sent out information to their specialist academic and other networks.

After six months of intensive recruitment activity the ICA Directorate and local advisors, which included a Fellow from ICA1, met across continents, remotely, to select the fellows. The committee carefully considered the suitability of the candidates based on statements they submitted on a potential project linked to the theme, academic performance and potential, past experiences, recommendations and area of expertise. Gender and geographical factors were also considered.

6. The Programme

"We had all the conditions we needed to make this a success and produce products and we will!" — ICA Fellow.

Planning of the intense and complex 10-day programme is embarked upon as soon as possible and begins with the availability of the Mentor Group. Once their availability is confirmed the organising group can plan the structure and timings of the sessions.

The building blocks for the programme:

- Mentor Talks and Responses
  Chaired by one of the ICA Directorate, the Mentor Talk is followed by a response and discussion from another mentor and two fellows. The fellows were allocated at random.
- 1:1s with Mentors
  Time was set aside after dinner for fellows to 1:1 or small group discussions with mentors.
- Thematic Discussion Panels
  Planned and organised by fellows on the Laws Theme with input from mentors.
- Project development by the Fellows
  Fellows were able to spend time planning writing and funding opportunities in thematic groups that they had established.
- Cultural and Social Programme
  In addition to living and working together the extensive additional programme introduced, food, dancing, history and culture and gave the group a unique insight into the host's setting and place.

  In Singapore, special trips were made to museums downtown and a small island, Pulau Ubin, of the eastern coast of Singapore. Pulau Ubin is not representative of modern-day Singapore and preserves traces of what Singapore looks like half a century ago. In UK, the participants were treated to an afternoon of Shakespearian culture and performance at Stratford-upon-Avon as well as an afternoon visiting museums and other monuments in downtown Birmingham.

Following feedback from fellows, the structure of the programme at Birmingham was organised to allow 12 sessions that the fellows organised and these they used to plan future research projects, and invite additional mentors to support particular ideas under development.

Additional elements in the Birmingham programme, were opportunities to network with other UK based researchers and extensive public programme with 12 very well attended public lectures from our mentors. It was extraordinary by any standards to have three Nobel Laureates on campus over the space of a week or so and the University's

Senior Management team were delighted to host several events that put Birmingham on the map! This is a unique opportunity linked to the ICA and demonstrates the reputational and networking possibilities that the gathering of eminent and well-connected researchers can bring to the host institution.

Facilities at both Singapore and Birmingham allowed for the flexible meeting spaces on the same site as the accommodation where, snacks refreshments and meals were available.

The layout of the rooms was adaptable with a set up conducive to discussions, also allowing for talks and panels. PowerPoint and recording facilities were available.

7. Interim Meetings

To support the momentum of nascent projects from Singapore, interim meetings were essential and the first step to this was identifying a location and host. Fellows were asked to submit a proposal for two-day meetings with a rationale, draft agenda and an indication of any financial support required for travel.

In October 2018, interim meetings were hosted at world class interdisciplinary institutes by mentors Patrick Geary at IAS Princeton and Ernst Rank, Institute for Advanced Study, Technical University of Munich.

The topics for the workshops "Changing Laws — Laws and Sustainability, Smart Cities, Dumb Laws" and "Deviating from Laws".

The majority of fellows were part of the two groups, hosting costs were met by IAS Princeton and IAS TUM and travel expenses by the Fellows' host institutes. The meeting at Princeton was also attended by the Director of the UBIAS network, Professor Ary Plonski to begin discussions on the establishment of an association of past fellows involved in the ICA, which would track career paths of past fellows, supporting interdisciplinary research projects, and planning ICA 4.

8. Passing on the Baton, Improvement and Continuity

Throughout the process the administrative team asked for and documented feedback from Fellows and Mentors to evaluate and improve the ICA concept to pass the baton on to the next ICA. After the Singapore and Birmingham meetings anonymous questionnaires allowed fellows to comment on the academic, cultural and organisational aspects of the programme. Directors Professor Rabinovici and Hannon also conducted individual sessions with fellows to discuss the programme and future academic trajectories. All this was collated and delivered to the ICA 4 team and Sue Gilligan joined the ICA 4 Steering Committee.

As in all interdisciplinary projects, discussion often arose from differences in the emphasis and understanding of various terms in each field. Discussions were often heated at the beginning but eventually acquired a steady state of mutual reconciliation, although even at the very last day of the second leg, it was not always clear who would like to lead a certain project, knowing how busy each participant will be after the face-to-face meeting.

Interdisciplinary collaboration is increasingly becoming a norm in frontier research nowadays. Yet there are perceived barriers to interdisciplinary research, especially among early career researchers. Moreover, there exists an optimal level of interdisciplinary research for maximal research impact: too many interdisciplinary contributions lose focus and niche, while too little interdisciplinary approach provides too much focus so that the research becomes too mainstream for significant effect outside the community.

The ICA does not have a fixed format: it relies on spontaneous assembly of people with some common interest including personality sometimes to identify a common problem and solve it. It almost allows Nature to take its own course. Identifying a common problem of interest sufficiently early to all self-assembled group members is therefore crucial to the success of the program.